W9-CID-435

ANTI-INVARIANT SUBMANIFOLDS

PURE AND APPLIED MATHEMATICS

A Program of Monographs, Textbooks, and Lecture Notes

Executive Editors — Monographs, Textbooks, and Lecture Notes

Earl J. Taft
Rutgers University
New Brunswick, New Jersey

Edwin Hewitt
University of Washington
Seattle, Washington

Chairman of the Editorial Board

S. Kobayashi
University of California, Berkeley
Berkeley, California

Editorial Board

Masanao Aoki
University of California, Los Angeles

Glen E. Bredon
Rutgers University

Sigurdur Helgason
Massachusetts Institute of Technology

G. Leitman
University of California, Berkeley

W.S. Massey
Yale University

Irving Reiner
University of Illinois at Urbana-Champaign

Paul J. Sally, Jr.
University of Chicago

Jane Cronin Scanlon
Rutgers University

Martin Schechter
Yeshiva University

Julius L. Shaneson
Rutgers University

LECTURE NOTES IN PURE AND APPLIED MATHEMATICS

1. *N. Jacobson,* Exceptional Lie Algebras
2. *L.-Å. Lindahl and F. Poulsen,* Thin Sets in Harmonic Analysis
3. *I. Satake,* Classification Theory of Semi-Simple Algebraic Groups
4. *F. Hirzebruch, W. D. Newmann, and S. S. Koh,* Differentiable Manifolds and Quadratic Forms
5. *I. Chavel,* Riemannian Symmetric Spaces of Rank One
6. *R. B. Burckel,* Characterization of C(X) Among Its Subalgebras
7. *B. R. McDonald, A. R. Magid, and K. C. Smith,* Ring Theory: Proceedings of the Oklahoma Conference
8. *Yum-Tong Siu,* Techniques of Extension of Analytic Objects
9. *S. R. Caradus, W. E. Pfaffenberger, and Bertram Yood,* Calkin Algebras and Algebras of Operators on Banach Spaces
10. *Emilio O. Roxin, Pan-Tai Liu, and Robert L. Sternberg,* Differential Games and Control Theory
11. *Morris Orzech and Charles Small,* The Brauer Group of Commutative Rings
12. *S. Thomeier,* Topology and Its Applications
13. *Jorge M. López and Kenneth A. Ross,* Sidon Sets
14. *W. W. Comfort and S. Negrepontis,* Continuous Pseudometrics
15. *Kelly McKennon and Jack M. Robertson,* Locally Convex Spaces
16. *M. Carmeli and S. Malin,* Representations of the Rotation and Lorentz Groups: An Introduction
17. *George B. Seligman,* Rational Methods in Lie Algebras
18. *Djairo Guedes de Figueiredo,* Functional Analysis: Proceedings of the Brazilian Mathematical Society Symposium
19. *Lamberto Cesari, Rangachary Kannan, and Jerry D. Schuur,* Nonlinear Functional Analysis of Differential Equations: Proceedings of the Michigan State University Conference
20. *Juan Jorge Schaffer,* Geometry of Spheres in Normed Spaces
21. *Kentaro Yano and Masahiro Kon,* Anti-Invariant Submanifolds

Other volumes in preparation

ANTI-INVARIANT SUBMANIFOLDS

Kentaro Yano
Tokyo Institute of Technology
Tokyo, Japan

Masahiro Kon
Science University of Tokyo
Tokyo, Japan

MARCEL DEKKER, INC. New York and Basel

COPYRIGHT © 1976 by MARCEL DEKKER, INC. ALL RIGHTS RESERVED.

Neither this book nor any part may be reproduced or transmitted in any form or by any means, electronic or mechanical, including photocopying, microfilming, and recording, or by any information storage and retrieval system, without permission in writing from the publisher.

MARCEL DEKKER, INC.

270 Madison Avenue, New York, New York 10016

LIBRARY OF CONGRESS CATALOG CARD NUMBER: 76-44562

ISBN: 0-8247-6555-9

Current printing (last digit):
10 9 8 7 6 5 4 3 2 1

PRINTED IN THE UNITED STATES OF AMERICA

1549759

INTRODUCTION

The study of anti-invariant submanifolds of Kaehlerian and Sasakian manifolds started just a few years ago. The purpose of these lecture notes is to gather and arrange the results on anti-invariant submanifolds of Kaehlerian and Sasakian manifolds obtained up to now and to invite the readers to the study of this very interesting topic of modern differential geometry.

In Chapter I we first of all recall the fundamental concepts of Riemannian, Kaehlerian, and Sasakian manifolds and in Chapter II those of submanifolds of these manifolds.

In Chapter III we study anti-invariant submanifolds of Kaehlerian manifolds.

Chapter IV is devoted to the study of anti-invariant submanifolds of Sasakian manifolds tangent to the structure vector field and Chapter V to the study of those normal to the structure vector field.

In Chapter VI we study some relations between submanifolds of Kaehlerian manifolds and those of Sasakian manifolds. For this purpose we use the method of Riemannian fibre bundles.

The authors wish to express here their deep gratitude to Professor S. Kobayashi who suggested that we include this book in the series "Lecture Notes in Pure and Applied Mathematics." It is a pleasant duty for us to acknowledge that Marcel Dekker, Inc. took all possible care in production of the book.

July 24, 1976

Kentaro Yano

Masahiro Kon

CONTENTS

Chapter I

Riemannian, Kaehlerian and Sasakian manifolds

In this chapter we state necessary preliminaries on Riemannian, Kaehlerian and Sasakian manifolds. For basic concepts of Riemannian manifolds we refer to Kobayashi-Nomizu [21] and Yano [68]. General theory of Kaehlerian manifolds may be found in Goldberg [13], Kobayashi-Nomizu [21, Vol. II] and Yano [67]. For Sasakian manifolds we refer to Sasaki [50].

§1. Riemannian manifolds

We consider an n-dimensional connected differentiable manifold M of class C^∞ covered by a system of coordinate neighborhoods $\{U;x^h\}$, where U denotes a neighborhood and x^h local coordinates in U, indices h, i, j,... taking values in the range $\{1,2,\ldots,n\}$.

If, from any system of coordinate neighborhoods covering the manifold M, we can choose a finite number of coordinate neighborhoods which covers the whole manifold, then M is said to be compact.

If we can cover the manifold M by a system of coordinate neighborhoods in such a way that the Jacobian determinant $|\partial x^{h'}/\partial x^h|$ of the coordinate transformation

$$x^{h'} = x^{h'}(x^1,\ldots,x^n)$$

in every non-empty intersection of two coordinate neighborhoods $\{U;x^h\}$ and $\{U';x^{h'}\}$ is always positive, then the manifold M is said to be orientable.

We denote by $\mathfrak{X}(M)$ the set of all vector fields on M. We now define a Riemannian metric on M. It is a tensor field g of type (0,2) which satisfies the following two conditions:

(i) it is symmetric: $g(X,Y) = g(Y,X)$ for any $X,Y \in \mathfrak{X}(M)$,

(ii) it is positive-definite: $g(X,X) \geq 0$ for every $X \in \mathfrak{X}(M)$ and $g(X,X) = 0$ if and only if $X = 0$.

A manifold M with Riemannian metric g is called a Riemannian manifold. A Riemannian metric gives rise to an inner product on each tangent space $T_x(M)$ to M at x. The components g_{ij} of g with respect to $\{U;x^h\}$ are given by

$$g_{ij} = g(\partial/\partial x^i, \partial/\partial x^j),$$

where $\partial/\partial x^i$ are base vectors of the so-called natural frame on the coordinate neighborhoods $\{U;x^h\}$. We call g_{ij} the covariant components of g. The contravariant component g^{ij} of g are defined by

$$g_{ij}g^{jk} = \delta_i^k, \qquad \delta_i^k = \begin{cases} 1, & k = i \\ 0, & k \neq i \end{cases},$$

where, here and in the sequel, we use the Einstein convention, that is, repeated indices, one upper index and the other lower index, denotes summation over its range. If X^i are components of a vector or a vector field X with respect to $\{U;x^h\}$, that is, $X = X^i(\partial/\partial x^i)$, then the components X_i of the corresponding covector or the corresponding 1-form are related to X^i by

$$X^i = g^{ij}X_j, \qquad X_i = g_{ij}X^j.$$

The inner product g in the tangent space $T_x(M)$ and in its dual space $T_x^*(M)$

can be extended to an inner product, denoted also by g, in the tensor space T^r_s at x for each type (r,s). If K and L are tensors at x of type (r,s) with components $K^{i_1\dots i_r}_{j_1\dots j_s}$ and $L^{i_1\dots i_r}_{j_1\dots j_s}$ (with respect to $\{U;x^h\}$), then the inner product g(K,L) of K and L is defined to be

$$g(K,L) = g_{i_1k_1}\cdots g_{i_rk_r}g^{j_1t_1}\cdots g^{j_st_s}K^{i_1\dots i_r}_{j_1\dots j_s}L^{k_1\dots k_r}_{t_1\dots t_s}.$$

When a Riemannian manifold M is orientable, we can define the volume element of M by

$$*1 = \sqrt{\mathfrak{g}}\,dx^1\wedge dx^2 \;\dots\wedge dx^n,$$

where $\mathfrak{g} = |g_{ij}|$ and $\wedge$ denotes the exterior product, and we can define the integral

$$\int_D f(x)*1$$

of a function f over the domain D of M.

Let M and N be Riemannian manifolds with Riemannian metrics g and h respectively. A mapping $f:M \longrightarrow N$ is said to be isometric at a point x of M if $g(X,Y) = h(f_*X,f_*Y)$ for all $X,Y\in T_x(M)$, where f_* is the differential of f. In this case, f_* is injective at x of M, because $f_*X = 0$ implies that $g(X,Y) = 0$ for all Y and hence $X = 0$. A mapping f which is isometric at every point of M is thus an immersion, which we call an isometric immersion. If, moreover, f is 1 : 1, then f is called an isometric imbedding of M into N.

§2. Covariant differentiation

An affine connection on a manifold M is a rule ∇ which assigns to each $X\in\mathfrak{X}(M)$ a linear mapping ∇_X of $\mathfrak{X}(M)$ into itself satisfying the following two conditions:

(∇_I) $\nabla_{fX+gY} = f\nabla_X + g\nabla_Y$,

(∇_{II}) $\nabla_X(fY) = f\nabla_X Y + (Xf)Y$

for $f,g \in \mathfrak{F}(M)$, $X,Y \in \mathfrak{X}(M)$, where $\mathfrak{F}(M)$ denotes the set of all differentiable functions on M.

The operator ∇_X is called the <u>covariant differentiation</u> with respect to X.

<u>Theorem 2.1</u>. On a Riemannian manifold M there exists one and only one affine connection satisfying the following two conditions:

(i) The torsion tensor T vanishes, i.e.,

$T(X,Y) = \nabla_X Y - \nabla_Y X - [X,Y] = 0$,

(ii) g is parallel, i.e., $\nabla_X g = 0$.

<u>Proof</u>. <u>Existence</u>. Given vector fields X and Y on M, we define $\nabla_X Y$ by setting

$$(2.1) \qquad 2g(\nabla_X Y,Z) = Xg(Y,Z) + Yg(X,Z) - Zg(X,Y) + g([X,Y],Z) + g([Z,X],Y) + g(X,[Z,Y]),$$

for any vector field Z on M. Then the mapping $(X,Y) \longrightarrow \nabla_X Y$ satisfies the condition (∇_I) and (∇_{II}), which shows that ∇ defines an affine connection on M. From the above definition of $\nabla_X Y$, we have $T(X,Y) = 0$ and

$$Xg(Y,Z) = g(\nabla_X Y,Z) + g(Y,\nabla_X Z),$$

which shows that $\nabla_X g = 0$.

<u>Uniqueness</u>. By a straightforward computation, we can see that, if

$\nabla_X Y$ satisfies $\nabla_X g = 0$ and $T(X,Y) = 0$, then it satisfies the equation which defines $\nabla_X Y$.

We define the function Γ^k_{ji} on $\{U;x^h\}$ by

$$\nabla_{\partial/\partial x^j}(\partial/\partial x^i) = \Gamma^k_{ji}(\partial/\partial x^k),$$

which are the components of the Riemannian connection. Putting $X = \partial/\partial x^j$, $Y = \partial/\partial x^i$ and $Z = \partial/\partial x^k$ in (2.1), we obtain

$$g_{lk}\Gamma^l_{ji} = \frac{1}{2}\left(\frac{\partial g_{ki}}{\partial x^j} + \frac{\partial g_{jk}}{\partial x^i} - \frac{\partial g_{ji}}{\partial x^k}\right).$$

Given a tensor field K of type (r,s), the <u>covariant derivative</u> $\nabla_X K$ of K with respect to X is defined to be

$$(\nabla_X K)(X_1,\ldots,X_s) = \nabla_X(K(X_1,\ldots,X_s)) - \sum_{i=1}^{s} K(X_1,\ldots,\nabla_X X_i,\ldots,X_s)$$

for vector fields X, $X_i \in \mathfrak{X}(M)$. The covariant derivative ∇K of K is a tensor field of type (r,s+1) defined by

$$(\nabla K)(X;X_1,\ldots,X_s) = (\nabla_X K)(X_1,\ldots,X_s).$$

A tensor field K on M is parallel, i.e., $\nabla_X K = 0$ for all $X \in T_x(M)$, if and only if $\nabla K = 0$.

We consider a p-form ω expressed as

$$\omega = \frac{1}{p!}\,\omega_{i_1 i_2\ldots i_p}\, dx^{i_1}\wedge dx^{i_2}\wedge\ldots\wedge dx^{i_p}$$

with respect to a system of local coordinates $x^1,\ldots,x^n$. The <u>exterior differential</u> $d\omega$ of ω is the (p+1)-form defined by

$$d\omega = \frac{1}{(p+1)!}(\partial_i \omega_{i_1 i_2 \ldots i_p} - \partial_{i_1}\omega_{i i_2 \ldots i_p}$$
$$- \ldots - \partial_{i_p}\omega_{i_1 i_2 \ldots i})dx^i \wedge dx^{i_1} \wedge \ldots \wedge dx^{i_p},$$

where we have put $\partial_i = \partial/\partial x^i$. The codifferential $\delta\omega$ of ω is the (p-1)-form defined by

$$\delta\omega = -\frac{1}{(p-1)!}(g^{ji}\nabla_j \omega_{i i_2 \ldots i_p})dx^{i_2} \wedge dx^{i_3} \wedge \ldots \wedge dx^{i_p}.$$

If f is a scalar, we put $\delta f = 0$. It can be verified that, for any p-form ω, we have

$$d^2\omega = 0, \qquad \delta^2\omega = 0.$$

A p-form ω is said to be harmonic if $d\omega = \delta\omega = 0$. If we put $\Delta = -\delta d - d\delta$, then we have $\Delta\omega = 0$ for a harmonic p-form ω. With a vector field X with local components X^h there is associated a 1-form γ given by

$$\gamma = g_{ji}X^j dx^i = X_i dx^i.$$

The codifferential $\delta\gamma$ of γ is given by

$$\delta\gamma = -\nabla_i X^i = -g^{ji}\nabla_j X_i.$$

We denote it by δX.

Theorem 2.2 (Green). Let X be a vector field on an orientable closed Riemannian manifold M. Then we have

$$\int_M (\delta X) * 1 = 0,$$

where a closed manifold means a compact manifold without boundary.

For a scalar function f, we obtain

$$\Delta f = g^{ji}\nabla_j\nabla_i f = \nabla^i\nabla_i f.$$

The differential operator $\Delta = g^{ji}\nabla_j\nabla_i$ or $\nabla^i\nabla_i$ is sometimes called the Laplacian. From Green's theorem, we have

Theorem 2.3. For any function f on an oriented closed Riemannian manifold M, we have

$$\int_M \Delta f * 1 = 0.$$

From this we have the following

Theorem 2.4 (Hopf). Let M be a closed Riemannian manifold. If f is a function on M such that $\Delta f \geq 0$ (or $\Delta f \leq 0$) everywhere, then f is a constant function.

Let M be a surface, that is, a 2-dimensional manifold, with Riemannian metric g, and let Δ be the Laplacian on M formed with g. A function f on M is called a subharmonic (resp. superharmonic) function on M if we have $\Delta f \geq 0$ (resp. $\Delta f \leq 0$) everywhere. A surface M is said to be parabolic if there exists no non-constant negative subharmonic function on M . Thus, if M is parabolic then every subharmonic function on M which is bounded from above on M must be a constant function on M.

§3. The structure equations of Cartan

Let M be an n-dimensional Riemannian manifold with the Riemannian connection ∇. We put

$$R(X,Y) = \nabla_X\nabla_Y - \nabla_Y\nabla_X - \nabla_{[X,Y]}$$

for all vector fields X and Y. R(X,Y) is a tensor field of type (1,1) which is linear in X and Y. We call R(X,Y) the <u>curvature transformation</u> on M. It can be easily verified that

$$R(X,Y) + R(Y,X) = 0,$$

(3.1)
$$R(X,Y)Z + R(Y,Z)X + R(Z,X)Y = 0.$$

Let $X_1,\ldots,X_n$ be a basis for the vector fields in some open neighborhood. We define the functions Γ^k_{ij}, R^i_{jkl} by

(3.2) $$\nabla_{X_i} X_j = \Gamma^k_{ij} X_k, \qquad R(X_k,X_l)X_j = R^i_{jkl} X_i,$$

respectively.

Let ω^i, ω^i_j be the 1-forms on M determined by

(3.3) $$\omega^i(X_j) = \delta^i_j, \qquad \omega^i_j = \Gamma^i_{kj}\omega^k.$$

It is clear that the forms ω^i_j determine the function Γ^i_{kj} and hence the connection ∇.

<u>Theorem 3.1 (the structure equations of Cartan)</u>.

(i) $$d\omega^i = -\omega^i_j \wedge \omega^j,$$

(ii) $$d\omega^i_j = -\omega^i_k \wedge \omega^k_j + \frac{1}{2} R^i_{jkl}\omega^k \wedge \omega^l.$$

<u>Proof</u>. If we define the functions c^i_{jk} by $[X_j,X_k] = c^i_{jk}X_i$, we obtain

$$d\omega^i(X_j,X_k) = \frac{1}{2}\{X_j\omega^i(X_k) - X_k\omega^i(X_j) - \omega^i([X_j,X_k])\} = -\frac{1}{2}c^i_{jk}.$$

Since the torsion tensor field T of M vanishes, we obtain

$$c^i_{jk}X_i = [X_j,X_k] = \nabla_{X_j}X_k - \nabla_{X_k}X_j = (\Gamma^i_{jk} - \Gamma^i_{kj})X_i,$$

so we have

$$c^i_{jk} = \Gamma^i_{jk} - \Gamma^i_{kj}.$$

On the other hand, we see that

$$-\omega^i_l \wedge \omega^l(X_j, X_k) = -\frac{1}{2}\{\omega^i_l(X_j)\omega^l(X_k) - \omega^l(X_j)\omega^i_l(X_k)\} = \frac{1}{2}(\Gamma^i_{kj} - \Gamma^i_{jk}).$$

From these equations we have (i). Similarly, we find

$$R^i_{jkl} = (\Gamma^p_{lj}\Gamma^i_{kp} - \Gamma^p_{kj}\Gamma^i_{lp}) + X_k\Gamma^i_{lj} - X_l\Gamma^i_{kj} - c^p_{kl}\Gamma^i_{pj},$$

$$d\omega^i_j(X_k, X_l) = \frac{1}{2}(X_k\Gamma^i_{lj} - X_l\Gamma^i_{kj} - c^p_{kl}\Gamma^i_{pj}),$$

$$-\omega^i_p \wedge \omega^p_j(X_k, X_l) = -\frac{1}{2}(\Gamma^p_{lj}\Gamma^i_{kp} - \Gamma^p_{kj}\Gamma^i_{lp}).$$

From these equations we have (ii).

§4. Curvature tensors

Let M be an n-dimensional Riemannian manifold. The Riemannian curva-tensor field R of M is the tensor field of covariant degree 4 defined by

$$(4.1) \qquad R(X_1,X_2,X_3,X_4) = g(R(X_3,X_4)X_2,X_1), \quad X_i \in T_x(M), \quad i = 1,\dots,4.$$

Then the Riemannian curvature tensor R can be considered as a quadrilinear mapping $T_x(M)\times T_x(M)\times T_x(M)\times T_x(M) \longrightarrow R$ at each point x of M and satisfies the following equations:

$$(4.2) \quad \begin{aligned} R(X_1,X_2,X_3,X_4) &= -R(X_2,X_1,X_3,X_4), \\ R(X_1,X_2,X_3,X_4) &= -R(X_1,X_2,X_4,X_3), \\ R(X_1,X_2,X_3,X_4) + R(X_1,X_3,X_4,X_2) &+ R(X_1,X_4,X_2,X_3) = 0, \\ R(X_1,X_2,X_3,X_4) &= R(X_3,X_4,X_1,X_2). \end{aligned}$$

If R^i_{jkl} and g_{ij} are components of the curvature tensor and the metric tensor with respect to a local coordinate system, then components R_{ijkl} of the Riemannian curvature tensor are given by $R_{ijkl} = g_{im}R^m_{jkl}$. Then the equations (4.2) can be expressed as

$$R_{ijkl} + R_{jikl} = 0, \qquad R_{ijkl} + R_{ijlk} = 0,$$

$$R_{ijkl} + R_{iklj} + R_{iljk} = 0, \qquad R_{ijkl} - R_{klij} = 0.$$

If $E_1,\ldots,E_n$ are local orthonormal vector fields, then

$$S(X,Y) = \sum_{i=1}^{n} g(R(E_i,X)Y,E_i) \tag{4.3}$$

defines a global tensor field S of type (0,2) with local components

$$R_{jl} = R^i_{jil} = g^{ik}R_{kjil}.$$

From the tensor field S we define a global scalar field

$$r = \sum_{i=1}^{n} S(E_i,E_i) \tag{4.4}$$

with local components

$$r = g^{ij}R_{ij}.$$

The tensor field S and the scalar function r are called the <u>Ricci tensor</u> and the <u>scalar curvature</u> of M respectively. If n = 2, then $G = \frac{1}{2}r$ is called the <u>Gaussian curvature</u>.

For each plane p in the tangent space $T_x(M)$ spanned by orthonormal vectors X_1 and X_2, the <u>sectional curvature</u> K(p) for p is defined by

$$K(p) = R(X_1,X_2,X_1,X_2) = g(R(X_1,X_2)X_2,X_1). \tag{4.5}$$

The sectional curvature K(p) is independent of the choice of an orthonormal basis X_1, X_2 and the set of values of K(p) for all plane p in $T_x(M)$ determines the Riemannian curvature tensor at x of M. If K(p) is a constant for all planes p in $T_x(M)$ and for all point x of M, then M is called a space of constant curvature. The following theorem due to Schur is well known (cf. [21, Vol. I]).

Theorem 4.1 (Schur). Let M be a connected Riemannian manifold of dimension $n > 2$. If the sectional curvature K(p) depends only on the point x, then M is a space of constant curvature.

A Riemannian manifold of constant curvature is called a space form. Sometimes, a space form is defined as a complete simply connected Riemannian manifold of constant curvature. We denote by M(c) a space form of constant curvature c. If M is a space of constant curvature c, then we have

$$R(X,Y)Z = c(g(Y,Z)X - g(X,Z)Y), \tag{4.6}$$

with its components

$$R_{ijkl} = c(g_{ik}g_{jl} - g_{il}g_{jk}) \quad \text{or} \quad R^i_{jkl} = c(\delta^i_k g_{jl} - \delta^i_l g_{jk}).$$

If we take an orthonormal frame field, then we have $g_{ij} = \delta_{ij}$ at x of M, and hence

$$R^i_{jkl} = R_{ijkl} = c(\delta_{ik}\delta_{jl} - \delta_{il}\delta_{jk}).$$

If the Ricci tensor S is of the form

$$S = \alpha g, \qquad R_{ij} = \alpha g_{ij},$$

M is called an Einstein space. If $n > 2$, we can see that α is a constant. If the curvature tensor R vanishes, that is, M is a space of zero curvature, then we call such a Riemannian manifold M a locally flat space. A Riemannian manifold M is called a locally symmetric space if its curvature tensor is parallel, that is, $\nabla R = 0$. A complete locally symmetric space is called a symmetric space.

The Weyl conformal curvature tensor field of M is the tensor field C of type (1.3) defined by

$$(4.7) \qquad C(X,Y)Z = R(X,Y)Z + \frac{1}{n-2}\{S(X,Z)Y - S(Y,Z)X + g(X,Z)QY - g(Y,Z)QX\} - \frac{r}{(n-1)(n-2)}\{g(X,Z)Y - g(Y,Z)X\}$$

for all vector fields X, Y and Z, where Q denotes the Ricci operator defined by $g(QX,Y) = S(X,Y)$. Moreover, we put

$$(4.8) \qquad c(X,Y) = (\nabla_X Q)Y - (\nabla_Y Q)X - \frac{1}{2(n-2)}\{(\nabla_X r)Y - (\nabla_Y r)X\}.$$

The tensor field C of type (1,3) vanishes identically for $n = 3$. Let ρ be a positive function on M. Then $g^* = \rho^2 g$ defines a change of metric on M which does not change the angle between two vectors at a point. Hence it is a conformal change of the metric. If a Riemannian metric g is conformally related to a Riemannian metric g^* which is locally flat, then the Riemannian manifold with the metric g is said to be conformally flat. The Weyl conformal curvature tensor C is invariant under any conformal change of the metric. The following is a well known theorem of Weyl.

Theorem 4.2. A necessary and sufficient condition for a Riemannian manifold M to be conformally flat is that $C = 0$ for $n > 3$ and $c = 0$ for $n = 3$.

It should be noted that if M is conformally flat and of dimension > 3, then $C = 0$ implies $c = 0$.

§5. Distribution

A q-dimensional <u>distribution</u> on an n-dimensional manifold M is a mapping D defined on M which assigns to each point x of M a q-dimensional linear subspace D_x of $T_x(M)$. A q-dimensional distribution D is said to be differentiable if there exist q differentiable vector fields in a neighborhood of x, forming a basis of D_y for each point y in this neighborhood of x. The set of these q vector fields is called a local <u>basis</u> of D. A vector field X belongs to D if $X_x \in D_x$ for any point x of M. We denote this fact by $X \in D$. A distribution D is said to be <u>involutive</u> if, for all vector fields X, Y in D, we have $[X,Y] \in D$. By a distribution we shall always mean a differentiable distribution.

A submanifold N imbedded in M is called an <u>integral manifold</u> of the distribution D if $f_*(T_x(M)) = D_x$ for all $x \in N$, where f_* is the differential of the imbedding f of N into M. If there exists no integral manifold of D which contains N, then N is called a <u>maximal integral submanifold</u> of D. A distribution D is said to be <u>completely integrable</u> if, for every point x of M, there is a unique integral manifold of D containing x.

The classical theorem of Frobenius can be stated as follows.

<u>Theorem 5.1 (Frobenius)</u>. An involutive distribution D on M is integrable. Moreover, through every $x \in M$ there passes a unique maximal integral manifold of D and every other integral manifold containing x is an open submanifold of this maximal one.

<u>Theorem 5.2</u>. A q-dimensional distribution D on M defined by a system of Pfaffian equations

$$\omega^1 = 0, \quad \omega^2 = 0, \ldots\ldots\ldots\ldots, \omega^q = 0$$

is completely integrable if and only if

$$d\omega^1 = 0, \quad d\omega^2 = 0, \ldots\ldots\ldots, d\omega^q = 0$$

with respect to any vector field of D.

§6. Kaehlerian manifolds

Let M be a real differentiable manifold. An almost complex structure on M is a tensor field J of type (1,1) which is, at every point x of M, an endomorphism of the tangent space $T_x(M)$ such that $J^2 = -I$, where I denotes the identity transformation of $T_x(M)$. A manifold M with a fixed almost complex structure J is called an almost complex manifold. Every almost complex manifold is of even dimensions and is orientable.

Let M be an almost complex manifold with almost complex structure J. We define the torsion of J to be the tensor field N of type (1,2), called the Nijenhuis tensor, given by

$$N(X,Y) = 2([JX,JY] - [X,Y] - J[X,JY] - J[JX,Y])$$

for all vector fields X and Y. If N vanishes identically, then an almost complex structure is called a complex structure and the manifold M is called a complex manifold.

A Hermitian metric on an almost complex manifold M is a Riemannian metric g such that

$$g(JX,JY) = g(X,Y) \qquad \text{for} \quad X,Y \in \mathfrak{X}(M).$$

An almost complex manifold (resp. a complex manifold) with Hermitian metric is called an almost Hermitian manifold (resp. a Hermitian manifold). Every almost complex manifold M with a Riemannian metric g admits a Hermitian metric. Indeed, setting

$$h(X,Y) = g(X,Y) + g(JX,JY) \qquad \text{for} \quad X,Y \in \mathfrak{X}(M),$$

we obtain a Hermitian metric h.

The fundamental 2-form Φ of an almost Hermitian manifold M is defined to be

$$\Phi(X,Y) = g(X,JY), \qquad X,Y \in \mathfrak{X}(M).$$

A Hermitian metric on an almost complex manifold is called a Kaehlerian metric if the fundamental 2-form is closed, i.e., $d\Phi = 0$. A complex manifold with a Kaehlerian metric is called a Kaehlerian manifold.

Let M be a Hermitian manifold with almost complex structure J and metric g. Let Φ be the fundamental 2-form and ∇ the operator of covariant differentiation defined by g. Then

$$2g((\nabla_X J)Y,Z) = 3d\Phi(X,JY,JZ) - 3d\Phi(X,Y,Z).$$

Consequently, the fundamental 2-form Φ is closed if and only if the almost complex structure J is parallel, that is, $\nabla_X J = 0$. Therefore a Hermitian manifold M is a Kaehlerian manifold if and only if the almost complex structure J of M is parallel.

The curvature tensor R of a Kaehlerian manifold has the following properties:

(6.1) $$R(X,Y)J = JR(X,Y) \quad \text{and} \quad R(JX,JY) = R(X,Y)$$

for all vector fields X and Y.

Let $K(p) = R(X,Y,X,Y)$ be the sectional curvature of a Kaehlerian manifold M for a plane p in $T_x(M)$ spanned by orthonormal vectors X and Y. If p is invariant by the complex structure J, then K(p) is called the holomorphic sectional curvature by p. If p is invariant by J and X is a unit vector in p, then X, JX is an orthonormal basis for p and hence $K(p) = R(X,JX,X,JX)$. If K(p) is a constant for all plane p in $T_x(M)$ invariant by J and for all points x of M, then M is called a space of constant holomorphic sectional curvature.

Theorem 6.1. Let M be a connected Kaehlerian manifold of complex dimension $n \geq 2$. If the holomorphic sectional curvature K(p) depends only on $x \in M$, then M is a space of constant holomorphic sectional curvature.

A Kaehlerian manifold of constant holomorphic sectional curvature is called a complex space form. Sometimes, a complex space form is assumed to be complete and simply connected. If M is of constant holomorphic sectional curvature c, then we have

$$(6.2) \qquad R(X,Y)Z = \frac{1}{4}c(g(Y,Z)X - g(X,Z)Y + g(Z,JY)JX - g(Z,JX)JY + 2g(X,JY)JZ),$$

with its local components

$$R^A_{BCD} = \frac{1}{4}c(\delta^A_C g_{BD} - \delta^A_D g_{BC} + J^A_C J_{BD} - J^A_D J_{BC} + 2J^A_B J_{CD}).$$

We denote by M(c) a complex space form of constant holomorphic sectional curvature c.

Let M be a complex n-dimensional Kaehlerian manifold. Then we can choose a local field of orthonormal frames $e_1,\ldots,e_n,e_{1*}=Je_1,\ldots,e_{n*}=Je_n$. We put

$$\omega^A(e_B) = \delta^A_B, \qquad \omega^A_B = \Gamma^A_{CD}\omega^C, \qquad \nabla_{e_A} e_B = \Gamma^C_{AB}e_C,$$

where A, B, C = 1,...,n,1*,...,n*. Then ω^A_B defines a Riemannian connection on M and we have

$$(6.3) \qquad \omega^i_j = \omega^{i*}_{j*}, \qquad \omega^{i*}_j = \omega^{j*}_i,$$

where i, j = 1,...,n.

Examples of Kaehlerian manifolds

1. The complex n-space C^n with the metric

$$ds^2 = \sum_{a=1}^{n} dz^a d\bar{z}^a,$$

where $\{z^1,\dots,z^n\}$ is the natural coordinate system, is a complete, flat Kaehlerian manifold with fundamental 2-form

$$\Phi = -i \sum_{a=1}^{n} dz^a \wedge d\bar{z}^a.$$

2. Let CP^n be a complex n-dimensional projective space with homogeneous coordinate system $\{z^0,z^1,\dots,z^n\}$. For each index j, let U_j be the open subset of CP^n defined by $z^j \neq 0$. We put

$$t_j^k = z^k/z^j, \qquad j, k = 0,\dots,n.$$

On each U_j, as a local coordinate system we take $t_j^0,\dots,\hat{t}_j^j,\dots,t_j^n$ (where $\hat{t}_j^j$ indicates that t_j^j is deleted) and consider the function f_j given by $f_j = \sum_{k=0}^{n} t_j^k \bar{t}_j^k$. Then

$$f_j = f_k t_j^k \bar{t}_j^k \qquad \text{on} \qquad U_j \cap U_k.$$

Since t_j^k is a holomorphic function in $U_j \cap U_k$, it follows that

$$\partial\bar{\partial}\log f_j = \partial\bar{\partial}\log f_k \qquad \text{on} \qquad U_j \cap U_k.$$

By putting

$$\Phi = -4i\partial\bar{\partial}\log f_j \qquad \text{on} \qquad U_j,$$

we obtain a globally defined (1,1)-form Φ on CP^n. We put

$$g(X,Y) = \Phi(JX,Y)$$

for all vector fields X and Y. To see that the symmetric form g is positive definite (so it is a Kaehlerian metric with fundamental form Φ), we verify this fact on each U_j by a direct calculation. On U_0 we have, for instance,

$$ds^2 = 4\frac{(1 + \Sigma_a t^a \bar{t}^a)(\Sigma_a dt^a d\bar{t}^a) - (\Sigma_a \bar{t}^a dt^a)(\Sigma_a t^a d\bar{t}^a)}{(1 + \Sigma_a t^a \bar{t}^a)^2},$$

where $t_0^a = t^a$, $a = 1,\ldots,n$. The Kaehlerian metric just constructed on CP^n is sometimes called the Fubini-Study metric.

Let S^{2n+1} be the unit sphere in C^{n+1} defined by $|z^0|^2 + \ldots + |z^n|^2 = 1$ and S^1 the multiplicative group of complex numbers of absolute value 1. Then S^{2n+1} is a principal fibre bundle over CP^n with group S^1.

From the following two theorems, a simply connected complete Kaehlerian manifold of constant holomorphic sectional curvature c can be identified with the complex projective space CP^n, the open unit ball D_n in C^n or C^n according as $c > 0$, $c < 0$ or $c = 0$.

Theorem 6.2. (1) For any positive number c, the complex projective space CP^n carries a Kaehlerian metric of constant holomorphic sectional curvature c. With respect to an inhomogeneous coordinate system $z^1,\ldots,z^n$, it is given by

$$ds^2 = \frac{4}{c}\,\frac{(1 + \Sigma z^a \bar{z}^a)(\Sigma dz^a d\bar{z}^a) - (\Sigma \bar{z}^a dz^a)(\Sigma z^a d\bar{z}^a)}{(1 + \Sigma z^a \bar{z}^a)^2}.$$

(2) For any negative number c, the open unit ball $D_n = \{(z^1,\ldots,z^n): z^a \bar{z}^a < 1\}$ in C^n carries a complex Kaehlerian metric of constant holomorphic sectional curvature c. With respect to the coordinate system $z^1,\ldots,z^n$ of C^n, it is given by

$$ds^2 = -\frac{4}{c}\frac{(1-\Sigma z^a\bar{z}^a)(\Sigma dz^a d\bar{z}^a) - (\Sigma\bar{z}^a dz^a)(\Sigma z^a d\bar{z}^a)}{(1-\Sigma z^a\bar{z}^a)^2}.$$

Theorem 6.3. Any two simply connected and complete Kaehlerian manifolds of constant holomorphic sectional curvature c are holomorphically isometric to each other.

§7. Sasakian manifolds

Let M be an odd-dimensional differentiable manifold of class C^∞ and ϕ, ξ, η be a tensor field of type (1,1), a vector field, a 1-form on M respectively such that

$$\phi^2 X = -X + \eta(X)\xi, \qquad \phi\xi = 0, \qquad \eta(\phi X) = 0, \qquad \eta(\xi) = 1$$

for any vector field X on M. Then M is said to have an almost contact structure (ϕ,ξ,η) and is called an almost contact manifold. The almost contact structure is said to be normal if

$$N + d\eta\otimes\xi = 0,$$

where N denotes the Nijenhius tensor formed with ϕ. If a Riemannian metric tensor field g is given on M and it satisfies the equations

$$g(\phi X,\phi Y) = g(X,Y) - \eta(X)\eta(Y), \qquad \eta(X) = g(X,\xi)$$

for any vector fields X and Y, then (ϕ,ξ,η,g)-structure is called an almost contact metric structure and we call M an almost contact metric manifold. If $d\eta(X,Y) = g(\phi X,Y)$ for any X, Y, then an almost contact metric structure is called a contact metric structure. If moreover the structure is normal, then a contact metric structure is called a Sasakian structure and a manifold M with Sasakian structure is called a Sasakian manifold. In a Sasakian manifold M with structure tensors (ϕ,ξ,η,g) we have

$$\nabla_X \xi = \phi X, \qquad (\nabla_X \phi)Y = -g(X,Y)\xi + \eta(Y)X$$

for any vector fields X and Y, where ∇ denotes the operator of covariant differentiation with respect to g. The Riemannian curvature tensor R of a Sasakian manifold M satisfies

$$R(X,\xi)Y = -g(X,Y)\xi + \eta(Y)X.$$

Moreover we have the following equations:

$$R(X,Y) = -\phi R(X,Y)\phi + X \wedge Y - \phi X \wedge \phi Y, \tag{7.1}$$

where $(X \wedge Y)Z = g(Y,Z)X - g(X,Z)Y$.

A plane section p in the tangent space $T_x(M)$ at x of a Sasakian manifold M is called a <u>ϕ-section</u> if it is spanned by a vector X orthogonal to ξ and ϕX. The sectional curvature K(p) with respect to a ϕ-section p determined by a vector X is called a <u>ϕ-sectional curvature</u>. It is verified that if a Sasakian manifold has a ϕ-sectional curvature c which does not depend on the ϕ-section at each point, then c is a constant in the manifold. A Sasakian manifold M is called a <u>Sasakian space form</u> and is denoted by M(c) if it has the constant ϕ-sectional curvature c. Sometimes, Sasakian space form is assumed to be simply connected and complete. The curvature tensor of a Sasakian space form M(c) is given by

$$\begin{aligned} R(X,Y)Z = {} & \frac{1}{4}(c+3)(g(Y,Z)X - g(X,Z)Y) - \frac{1}{4}(c-1)(\eta(Y)\eta(Z)X \\ & - \eta(X)\eta(Z)Y + g(Y,Z)\eta(X)\xi - g(X,Z)\eta(Y)\xi \\ & - g(\phi Y,Z)\phi X + g(\phi X,Z)\phi Y + 2g(\phi X,Y)\phi Z) \end{aligned} \tag{7.2}$$

and its local components by

$$R^A_{BCD} = \frac{1}{4}(c+3)(\delta^A_C g_{BD} - \delta^A_D g_{BC}) + \frac{1}{4}(c-1)(\eta_B\eta_C\delta^A_D - \eta_B\eta_D\delta^A_C$$
$$+ \eta^A\eta_D g_{BC} - \eta^A\eta_C g_{BD} + \phi^A_C\phi_{BD} - \phi^A_D\phi_{BC} + 2\phi^A_B\phi_{CD}).$$

Let M be a (2n+1)-dimensional Sasakian manifold. Then we can choose a local field of orthonormal frames $e_0=\xi,e_1,\ldots,e_n,e_{1*}=\phi e_1,\ldots,e_{n*}=\phi e_n$. With respect to this frame field, let $\omega^0=\eta,\omega^1,\ldots,\omega^n,\omega^{1*},\ldots,\omega^{n*}$ be its dual frames. Then we see that

$$(7.3) \qquad \omega^i_j = \omega^{i*}_{j*}, \qquad \omega^{i*}_j = \omega^{j*}_i, \qquad \omega^i = \omega^{i*}_0, \qquad \omega^{i*} = -\omega^i_0,$$

where i = 1,...,n.

Examples of Sasakian manifolds

1. Let S^{2n+1} be a (2n+1)-dimensional unit sphere, i.e.,

$$S^{2n+1} = \{z \in C^{n+1} : |z| = 1\}.$$

For any point $z \in S^{2n+1}$, put $\xi = Jz$, where J is the almost complex structure of C^{n+1}. We consider the orthogonal projection

$$\pi : T_z(C^{n+1}) \longrightarrow T_z(S^{2n+1}).$$

Putting $\phi = \pi \cdot J$, we have a Sasakian structure (ϕ,ξ,η,g) on S^{2n+1}, where η is a 1-form dual to ξ and g the standard metric tensor field on S^{2n+1}. We see that S^{2n+1} is of constant ϕ-sectional curvature 1, that is, of constant curvature 1.

2. Let E^{2n+1} be a Euclidean space with cartesian coordinates $(x^1,\ldots,x^n,$ $y^1,\ldots,y^n,z)$. Then the Sasakian structure on E^{2n+1} is defined by ϕ, ξ, η and g such that

$$\xi = (0,\ldots,0,2), \qquad 2\eta = (-y^1,\ldots,-y^n,0,\ldots,0,1),$$

$$
(g_{AB}) = \begin{pmatrix} \frac{1}{4}(\delta_{ij} + y^i y^j) & 0 & -\frac{1}{4}y^i \\ 0 & \frac{1}{4}\delta_{ij} & 0 \\ -\frac{1}{4}y^i & 0 & \frac{1}{4} \end{pmatrix},
$$

$$
(\phi^A_B) = \begin{pmatrix} 0 & \delta^i_j & 0 \\ -\delta^i_j & 0 & 0 \\ 0 & y^j & 0 \end{pmatrix}.
$$

Then E^{2n+1} is of constant ϕ-sectional curvature -3 and denoted by $E^{2n+1}(-3)$.

Chapter II

Submanifolds

In this chapter we summarize the basic results on submanifolds. For submanifolds of Riemannian manifolds we refer to Chen [5] and Kobayashi-Nomizu [21, Vol. II]. Fundamental concepts of Kaehlerian submanifolds are found in Ogiue [45] and those of Sasakian submanifolds in Kon [22, 28] and Yano-Ishihara [75].

§1. Induced connection and second fundamental form

Let M be an n-dimensional manifold immersed in an m-dimensional Riemannian manifold $\bar{M}$ where $m > n$. We denote by $\bar{\nabla}$ the operator of covariant differentiation in $\bar{M}$ and by $\bar{g}$ the Riemannian metric tensor field on $\bar{M}$. Since our discussion is local, we may assume that M is imbedded in $\bar{M}$. The induced metric on M is given by $g(X,Y) = \bar{g}(X,Y)$ for any vector fields X and Y on M. Let $T(M)$ and $T(M)^{\perp}$ denote the tangent and normal bundle of M respectively. The connection $\bar{\nabla}$ and the metric $\bar{g}$ lead to connections and invariant inner products on $T(M)$ and $T(M)^{\perp}$. The connection on $T(M)$ is the unique Riemannian connection induced by the inherited metric g. We will define these connections explicitely.

Let X and Y be vector fields on M. For each point x of M, we put

$$(1.1) \qquad (\nabla_X Y)_x = (\bar{\nabla}_X Y)^T_x,$$

where $(\ )^T$ denotes the projection into $T_x(M)$.

Proposition 1.1. ∇ is the unique Riemannian connection on $T(M)$ with respect to the induced metric on M.

Proof. By the definition of ∇, it is an affine connection on M. To show that ∇ has torsion 0, we compute

$$\nabla_X Y - \nabla_Y X - [X,Y] = (\bar{\nabla}_X Y)^T - (\bar{\nabla}_Y X)^T - [X,Y]^T$$
$$= (\bar{\nabla}_X Y - \bar{\nabla}_Y X - [X,Y])^T = 0.$$

Moreover, we see that

$$Xg(Y,Z) = X\bar{g}(Y,Z) = \bar{g}(\bar{\nabla}_X Y,Z) + \bar{g}(Y,\bar{\nabla}_X Z)$$
$$= g(\nabla_X Y,Z) + g(Y,\nabla_X Z),$$

which means $\nabla g = 0$. Thus ∇ is the Riemannian connection of $T(M)$, that is, that on M.

The connection in the normal bundle $T(M)^{\perp}$ is defined similarly. Let N be a normal vector field on M and X a tangent vector field of M. For each point x of M we put

$$(1.2) \qquad (D_X N)_x = (\bar{\nabla}_X N)^{\perp}_x,$$

where $(\ )^{\perp}$ denotes the projection into $T_x(M)^{\perp}$.

Proposition 1.2. D is a metric connection in the normal bundle $T(M)^{\perp}$ with respect to the induced metric on $T(M)^{\perp}$.

Proof. We easily see that D defines an affine connection on the normal bundle $T(M)^{\perp}$. For normal vector fields N and V, (1.2) implies that

$$X\bar{g}(N,V) = \bar{g}(\bar{\nabla}_X N,V) + \bar{g}(N,\bar{\nabla}_X V) = \bar{g}(D_X N,V) + \bar{g}(N,D_X V),$$

which shows that D is a metric connection.

We define two vector bundles over M associated with T(M) and $T(M)^{\perp}$. Let S(M) be the bundle whose fibre at each point is a space of symmetric linear transformations of $T_x(M) \longrightarrow T_x(M)$. Let us put $H(M) = \mathrm{Hom}(T(M)^{\perp}, S(M))$. For a normal vector field N on M and for a tangent vector field X on M, we put

$$(1.3) \qquad (A_N X)_x = -(\bar{\nabla}_X N)_x^T$$

at each point x of M. Then A is well defined, that is, $(A_N X)_x$ depends only on X_x and N_x. We obtain

$$g(A_N X,Y) - g(A_N Y,X) = -\bar{g}(\bar{\nabla}_X N,Y) + \bar{g}(\bar{\nabla}_Y N,X)$$

$$= \bar{g}(N,\bar{\nabla}_X Y) - \bar{g}(N,\bar{\nabla}_Y X) = \bar{g}(N,[X,Y]) = 0.$$

Consequently $A_N : T_x(M) \longrightarrow T_x(M)$ is a symmetric linear transformation. Clearly A is linear in all variable and therefore $A \in H(M)$. We call A the <u>second fundamental form</u> of M. It will sometimes be more convenient to regard the second fundamental form as a symmetric bilinear form on $T_x(M)$ with values in $T_x(M)^{\perp}$. That is, for X, $Y \in T_x(M)$, we define $B(X,Y) \in T_x(M)^{\perp}$ by

$$(1.4) \qquad \bar{g}(B(X,Y),N) = g(A_N X,Y).$$

<u>Proposition 1.3</u>. At each point x of M, we have

$$(1.5) \qquad B(X,Y)_x = (\bar{\nabla}_X Y)_x^{\perp}.$$

<u>Proof</u>. From (1.3) and (1.4) we see

$$\bar{g}(B(X,Y),N) = g(A_N X,Y) = -\bar{g}(\bar{\nabla}_X N,Y)$$

$$= -X\bar{g}(N,Y) + \bar{g}(N,\bar{\nabla}_X Y) = \bar{g}(\bar{\nabla}_X Y,N).$$

We have thus proved the first set of basic formulas for submanifolds, namely,

(I) $$\bar{\nabla}_X Y = \nabla_X Y + B(X,Y),$$

(II) $$\bar{\nabla}_X N = - A_N X + D_X N.$$

The equation (I) is called the Gauss formula and (II) the Weingarten formula.

In the next place we descrive the second fundamental form B in term of local coordinates.

Let x be a point of M and let $Y_1,\ldots,Y_n,Y_{n+1},\ldots,Y_m$ be an orthonormal basis of $T_x(\bar{M})$ such that $Y_1,\ldots,Y_n$ form a basis of $T_x(M)$. We may choose a system of normal coordinates $y^1,\ldots,y^m$ such that $(\partial/\partial y^A)_x = Y_A$, $1 \leq A \leq m$. Note that $Y_{n+1},\ldots,Y_m$ form a basis of $T_x(M)^\perp$. Let $\{x^1,\ldots,x^n\}$ be an arbitrary coordinate system in a neighborhood U of x in M and let

$$y^A = y^A(x^1,\ldots,x^n), \quad 1 \leq A \leq m,$$

be the system of equations that defines the imbedding of U into $\bar{M}$. We shall show that

$$B((\partial/\partial x^i)_x,(\partial/\partial x^j)_x) = \sum_{a=n+1}^{m} (\partial^2 y^a/\partial x^i \partial x^j)_x Y_a, \tag{1.6}$$

namely, the coefficients of B_x with respect to the basis $(\partial/\partial x^1)_x,\ldots,$ $(\partial/\partial x^n)_x$ in $T_x(M)$ and the basis $Y_{n+1},\ldots,Y_m$ in $T_x(M)^\perp$ are those of the Hessian $\partial^2 y^a/\partial x^i \partial x^j$ at x. Indeed, we have

$$\begin{aligned}\bar{\nabla}_{\partial/\partial x^i}(\partial/\partial x^j) &= \bar{\nabla}_{\partial/\partial x^i}\Big(\sum_{A=1}^{m} (\partial y^A/\partial x^j)(\partial/\partial y^A)\Big)\\ &= \sum_{A,B=1}^{m} (\partial y^A/\partial x^j)(\partial y^B/\partial x^i)\bar{\nabla}_{\partial/\partial y^B}(\partial/\partial y^A) + \sum_{A=1}^{m} (\partial^2 y^A/\partial x^i \partial x^j)(\partial/\partial y^A)\\ &= \sum_{A,B,C=1}^{m} (\partial y^A/\partial x^j)(\partial y^B/\partial x^i)\bar{\Gamma}^C_{AB}(\partial/\partial y^C) + \sum_{A=1}^{m} (\partial^2 y^A/\partial x^i \partial x^j)(\partial/\partial y^A),\end{aligned}$$

where $\bar{\Gamma}^C_{AB}$ are the Christoffel symbols for the Riemannian connection in $\bar{M}$ with respect to $y^1,\ldots,y^m$. Note that $\bar{\Gamma}^C_{AB}$ are 0 at the origin x of the normal coordinate system. Taking the normal components at x of both sides of the equation above, we have (1.6).

A normal vector field N on M is said to be <u>parallel in the normal bundle</u>, or simply <u>parallel</u>, if $D_X N = 0$ for all vector field X on M. A submanifold M is said to be <u>totally geodesic</u> if its second fundamental form vanishes identically, i.e., B = 0 (A = 0). For a normal section N on M, if A_N is everywhere proportional to the identity transformation I, i.e., $A_N = \alpha I$ for some function α, then N is called an <u>umbilical section</u> on M, or M is said to be umbilical with respect to N. If the submanifold M is umbilical with respect to every local normal section in M, then M is said to be <u>totally umbilical</u>. Let $e_1, \ldots, e_n$ be an orthonormal basis in $T_x(M)$. Then the <u>mean curvature vector</u> m of M is defined to be $m = (\mathrm{Tr}B)/n$, where $\mathrm{Tr}B = \sum_i B(e_i, e_i)$, which is independent of the choice of a basis. We obtain $m = \sum_a \frac{1}{n}(\mathrm{Tr}A_a)e_a$, where $\{e_a\}$, a = n+1,...,m, is an orthonormal basis in $T_x(M)^{\perp}$, and we denote A_{e_a} by A_a for the simplicity. If m = 0, M is said to be <u>minimal</u>.

§2. Gauss, Codazzi and Ricci equations

Let M be an n-dimensional manifold immersed in an m-dimensional Riemannian manifold $\bar{M}$. We denote by K the Riemannian curvature tensor of $\bar{M}$ and by R the Riemannian curvature tensor of M. For the second fundamental form B of M we define its covariant derivative $\nabla_X B$ by

$$(2.1) \qquad (\nabla_X B)(Y,Z) = D_X(B(Y,Z)) - B(\nabla_X Y, Z) - B(Y, \nabla_X Z)$$

for any vector fields X, Y and Z on M, which is defined equivalently by putting

$$(2.2) \qquad (\nabla_X A)_N Y = \nabla_X(A_N Y) - A_{D_X N} Y - A_N(\nabla_X Y)$$

for any normal vector field N to M. If $\nabla_X B = 0$ for all X, then the second fundamental form of M is said to be parallel. By (1.4), (2.1) and (2.2), the second fundamental form of M is parallel if and only if $\nabla_X A = 0$ for all X. By a straightforward computation, which uses Gauss and Weingarten formulas, we obtain

$$K(X,Y)Z = R(X,Y)Z - A_{B(Y,Z)}X + A_{B(X,Z)}Y + (\nabla_X B)(Y,Z) - (\nabla_Y B)(X,Z) \tag{2.3}$$

for any vector fields X, Y and Z on M. For any vector field W on M, by (1.4) and (2.3) we have the following Gauss equation:

$$\bar{g}(K(X,Y)Z,W) = g(R(X,Y)Z,W) - \bar{g}(B(X,W),B(Y,Z)) - \bar{g}(B(Y,W),B(X,Z)). \tag{2.4}$$

Taking the normal component of the equation (2.3), we obtain the Codazzi equation:

$$(K(X,Y)Z)^{\perp} = (\nabla_X B)(Y,Z) - (\nabla_Y B)(X,Z). \tag{2.5}$$

We define the curvature tensor $R^{\perp}$ of the normal connection D by

$$R^{\perp}(X,Y)N = D_X D_Y N - D_Y D_X N - D_{[X,Y]}N \tag{2.6}$$

for any vector fields X, Y on M and normal vector field N on M. From the Gauss and Weingarten formulas we see

$$\bar{g}(K(X,Y)N,V) = \bar{g}(R^{\perp}(X,Y)N,V) + g([A_V,A_N]X,Y) \tag{2.6}$$

for any normal vector field V, where $[A_V,A_N] = A_V A_N - A_N A_V$. Equation (2.6) is called the Ricci equation. If $R^{\perp} = 0$, then the normal connection of M is said to be flat. When $\bar{M}$ is of constant curvature, the normal connection of M is flat if and only if the second fundamental form of M is commutative, i.e., $[A_V, A_N] = 0$ for all V and N. If K(X,Y)Z is tangent to M, then the Codazzi equation (2.5) reduces to

$$(2.7) \qquad (\nabla_X B)(Y,Z) - (\nabla_Y B)(X,Z) = 0.$$

Sometimes, we call the equation (2.7) the Codazzi equation.

In the following, we call q-plane bundle over a manifold M a Riemannian q-plane bundle if it is equipped with a bundle metric and a compatible connection. If E is a q-plane bundle over a Riemannian manifold M, a second fundamental form in E is a section A of Hom(T(M)×E,T(M)) such that

$$g(A(X,N),Y) = g(X,A(Y,N))$$

for any vector fields X, Y on M and N in E. If E is a Riemannian vector bundle with second fundamental form A, we define its associated second fundamental form B taking values in E by

$$g(B(X,Y),N) = \bar{g}(A(X,N),Y),$$

where $\bar{g}$ denotes the bundle metric in E.

We can now state the Fundamental theorems of submanifolds as follows. For the proofs, see Bishop-Crittenden [2].

Theorem 2.1 (Existence theorem). Let M be a simply connected n-dimensional Riemannian manifold with a Riemannian q-plane bundle E over M eqipped with a second fundamental form A and associated second fundamental form B. If they satisfy the equations (2.4), (2.6) and (2.7), then M can be isometrically immersed in (n+q)-dimensional space form $\bar{M}(c)$ of curvature c with normal bundle E.

Theorem 2.2 (Rigidity theorem). Let $f,g : M \longrightarrow \bar{M}(c)$ be two isometric immersions of an n-dimensional Riemannian manifold M in an m-dimensional space form $\bar{M}(c)$ of curvature c with normal bundles E, E' equipped with their canonical bundle metrics, connections and second fundamental forms. Suppose that there is an isometry $h ; M \longrightarrow M$ such that h can be covered by a bundle map $\bar{h} : E \longrightarrow E'$ which preserves the bundle metrics, the connections and the second fundamental forms. Then there is a rigid motion F of $\bar{M}(c)$ such that $F \cdot f = g \cdot h$.

§3. Structure equations of submanifolds

Let M be an n-dimensional submanifold immersed in an m-dimensional Riemannian manifold $\bar{M}$. We choose a local field of orthonormal frames $e_1,\ldots,e_m$ in $\bar{M}$ in such a way that, restricted to M, the vectors $e_1,\ldots,e_n$ are tangent to M and hence $e_{n+1},\ldots,e_m$ are normal to M. With respect to this frame field of $\bar{M}$, let $\omega^1,\ldots,\omega^m$ be the field of dual frames. We shall make use of the following convention on the ranges of indices:

$$A,B,C,\ldots = 1,\ldots\ldots,m,$$

$$i,j,k,\ldots = 1,\ldots\ldots n,$$

$$a,b,c,\ldots = n+1,\ldots\ldots,m.$$

Then the structure equations of $\bar{M}$ are given by

$$(3.1) \qquad d\omega^A = -\omega^A_B \wedge \omega^B, \qquad \omega^A_B + \omega^B_A = 0,$$

$$(3.2) \qquad d\omega^A_B = -\omega^A_C \wedge \omega^C_B + \Phi^A_B, \qquad \Phi^A_B = \frac{1}{2}K^A_{BCD}\omega^C \wedge \omega^D.$$

We restrict thse forms to M. Then

$$(3.3) \qquad \omega^a = 0.$$

Since $0 = d\omega^a = -\omega_i^a \wedge \omega^i$, by Cartan's lemma, we obtain

$$(3.4) \qquad \omega_i^a = h_{ij}^a \omega^j, \qquad h_{ij}^a = h_{ji}^a.$$

From (3.2), (3.3) of Chapter I and (1.3) we see that $g(A_a e_i, e_j) = h_{ij}^a$, where we have put $A_a = A_{e_a}$. Thus h_{ij}^a are components of the second fundamental form A_a with respect to e_a. Therefore A_a can be considered as a symmetric (n,n)-matrix $A_a = (h_{ij}^a)$. Moreover we have the following equations:

$$(3.5) \qquad d\omega^i = -\omega_k^i \wedge \omega^k, \qquad \omega_k^i + \omega_i^k = 0,$$

$$(3.6) \qquad d\omega_j^i = -\omega_k^i \wedge \omega_j^k + \Omega_j^i, \qquad \Omega_j^i = \frac{1}{2} R_{jkl}^i \omega^k \wedge \omega^l,$$

$$(3.7) \qquad R_{jkl}^i = K_{jkl}^i + \sum_a (h_{ik}^a h_{jl}^a - h_{il}^a h_{jk}^a),$$

$$(3.8) \qquad d\omega_b^a = -\omega_c^a \wedge \omega_b^c + \Omega_b^a, \qquad \Omega_b^a = \frac{1}{2} R_{bkl}^a \omega^k \wedge \omega^l,$$

$$(3.9) \qquad R_{bkl}^a = K_{bkl}^a + \sum_i (h_{ik}^a h_{il}^b - h_{il}^a h_{ik}^b).$$

Equation (3.7) is the local expression of the Gauss equation (2.4) and equation (3.9) is the local expression of the Ricci equation (2.6). The forms (ω_j^i) define the Riemannian connection of M and the forms (ω_b^a) define the connection induced in the normal bundle of M. The second fundamental form of M is represented by $h_{ij}^a \omega^i \omega^j e_a$ and is sometimes denoted by its components h_{ij}^a. The second fundamental form of M is commutative if and only if $\sum_j (h_{ij}^a h_{jk}^b - h_{jk}^a h_{ij}^b) = 0$ for all a, b, i and k. If $R_{bkl}^a = 0$ for all a, b, k and l, then the normal connection of M is flat. The mean curvature vector m of M is given by $(\sum_k h_{kk}^a e_a)/n$ and if the second fundamental form of M is of the form $\delta_{ij}(\sum_k h_{kk}^a e_a)/n$, M is totally umbilical. If $\sum_k h_{kk}^a = 0$ for all a, then M is a minimal submanifold.

The covariant derivative h^a_{ijk} of h^a_{ij} is given by

$$(3.10) \qquad h^a_{ijk}\omega^k = dh^a_{ij} - h^a_{il}\omega^l_j - h^a_{lj}\omega^l_i + h^b_{ij}\omega^a_b.$$

If $h^a_{ijk} = 0$ for all indices, then the second fundamental form of M is parallel. The Laplacian Δh^a_{ij} of h^a_{ij} is defined as

$$(3.11) \qquad \Delta h^a_{ij} = \sum_k h^a_{ijkk},$$

where we have put

$$(3.12) \qquad h^a_{ijkl}\omega^l = dh^a_{ijk} - h^a_{ljk}\omega^l_i - h^a_{ilk}\omega^l_j - h^a_{ijl}\omega^l_k + h^b_{ijk}\omega^a_b.$$

Now we compute the Laplacian of the second fundamental form (see [10], [51] and [76]). In the following we assume that the second fundamental form of M satisfies the Codazzi equation (2.7), that is,

$$(3.13) \qquad h^a_{ijk} - h^a_{ikj} = 0.$$

From (3.12) we have

$$(3.14) \qquad h^a_{ijkl} - h^a_{ijlk} = h^a_{it}R^t_{jkl} + h^a_{tj}R^t_{ikl} - h^b_{ij}R^a_{bkl}.$$

On the other hand, (3.11) and (3.13) imply that

$$(3.15) \qquad \Delta h^a_{ij} = \sum_k h^a_{ijkk} = \sum_k h^a_{kijk}.$$

From (3.13), (3.14) and (3.15) we have

$$(3.16) \qquad \Delta h^a_{ij} = \sum_k (h^a_{kkij} + h^a_{kt}R^t_{ijk} + h^a_{ti}R^t_{kjk} - h^b_{ki}R^a_{bjk}).$$

Consequently, we obtain

$$(3.17)\qquad \sum_{a,i,j} h^a_{ij}\Delta h^a_{ij} = \sum_{a,i,j,k} (h^a_{ij}h^a_{kkij} + h^a_{ij}h^a_{kt}R^t_{ijk} + h^a_{ij}h^a_{ti}R^t_{kjk} - h^a_{ij}h^b_{ki}R^a_{bjk}).$$

By (3.7), (3.9) and (3.17) we have the following equation

$$(3.18)\qquad \sum_{a,i,j} h^a_{ij}\Delta h^a_{ij} = \sum_{a,i,j,k} (h^a_{ij}h^a_{kkij} + K^t_{ijk}h^a_{ij}h^a_{kt} + K^t_{kjk}h^a_{ij}h^a_{ti} - K^a_{bjk}h^a_{ij}h^a_{ki}) - \sum_{a,b,i,j,k,t} [(h^a_{ik}h^b_{kj} - h^a_{jk}h^b_{ki})(h^a_{it}h^b_{tj} - h^a_{jt}h^b_{ti}) + h^a_{ij}h^b_{ij}h^a_{tk}h^b_{tk} - h^a_{ij}h^b_{jt}h^a_{ti}h^b_{kk}].$$

§4. Kaehlerian submanifolds

Let $\bar{M}$ be a Kaehlerian manifold of complex dimension m (of real dimension 2m) with almost complex structure J and with Kaehlerian metric $\bar{g}$. Let M be a complex n-dimensional analytic submanifold of $\bar{M}$, that is the immersion $f : M \longrightarrow \bar{M}$ is holomorphic, i.e., $J\cdot f_* = f_*\cdot J$, where f_* is the differential of the immersion f and we denote by the same J the induced complex structure on M. Then the Riemannian metric g induced on M is Hermitian. It is easy to see that the fundamental 2-form with this Hermitian metric g is the restriction of the fundamental 2-form of $\bar{M}$ and hence is closed. This shows that every complex analytic submanifold M of a Kaehlerian manifold $\bar{M}$ is also a Kaehlerian manifold with the induced metric. We call such a submanifold M of $\bar{M}$ a Kaehlerian submanifold. In other words, a Kaehlerian submanifold M of a Kaehlerian manifold $\bar{M}$ is an invariant submanifold under the action of the complex structure J of $\bar{M}$, i.e., $JT_x(M) \subset T_x(M)$ for every point x of M. Then we see that $JT_x(M)^\perp \subset T_x(M)^\perp$ for every x of M. In the following, we denote by $\bar{\nabla}$ (resp. ∇) the operator of covariant differentiation with respect to $\bar{g}$ (resp. g). For any vector fields X and Y on M, the Gauss and Weingarten formulas imply that

$$\bar{\nabla}_X JY = J\nabla_X Y + B(X,JY), \qquad \bar{\nabla}_X JY = J\nabla_X Y + JB(X,Y),$$

which imply the following equations

$$B(X,JY) = B(JX,Y) = JB(X,Y). \tag{4.1}$$

Proposition 4.1. Any Kaehlerian submanifold M is a minimal submanifold.

Proof. For each $T_x(M)$ we can choose a basis $e_1,\ldots,e_n,Je_1,\ldots,Je_n$. Then (4.1) implies

$$\sum_{i=1}^{n} (B(e_i,e_i) + B(Je_i,Je_i)) = 0,$$

which means that M is a minimal submanifold.

From the Gauss equation (2.4) and (4.1) we have

Proposition 4.2. Let M be a Kaehlerian submanifold of a Kaehlerian manifold $\bar{M}$ and let R and K be the Riemannian curvature tensors of M and $\bar{M}$ respectively. Then

$$R(X,JX,X,JX) = K(X,JX,X,JX) - 2\bar{g}(B(X,X),B(X,X))$$

for all vector fields X on M.

As a direct consequence of Proposition 4.2, we obtain

Proposition 4.3. Any Kaehlerian submanifold M of a complex space form $\bar{M}(c)$ of constant holomorphic sectional curvature c is totally geodesic if and only if M is of constant holomorphic sectional curvature c.

Example 4.1. Let $CP^m(c)$ be a complex projective space of complex dimension m and of constant holomorphic sectional curvature $c \geq 0$. Any complex projective space $CP^n(c)$ $(m > n)$ is totally geodesic in $CP^m(c)$.

Example 4.2. Let $CP^m(c)$ be a complex projective space in Example 4.1. Let $z^0, z^1, \ldots, z^m$ be a homogeneous coordinate system of $CP^m(c)$. Let

$$Q^{m-1} = \{(z^0, z^1, \ldots, z^m) \in CP^m(c) : \sum_{i=0}^{m} (z^i)^2 = 0\}.$$

Then Q^{m-1} is complex analytically isometric to the Hermitian symmetric space $SO(m+1)/SO(2) \times SO(m-1)$. Q^{m-1} is called a hyperquadric of $CP^m(c)$. The hyperquadric Q^{m-1} is an Einstein Kaehlerian submanifold of $CP^m(c)$.

Example 4.3. The complex projective space $CP^n(c)$ is a Kaehlerian submanifold of the complex projective space $CP^{n+p}(2c)$, where $p = n(n+1)/2$.

§5. Anti-invariant submanifolds of complex manifolds

Let $\bar{M}$ be a real 2m-dimensional (complex m-dimensional) almost Hermitian manifold with almost complex structure J and with Hermitian metric $\bar{g}$. An n-dimensional Riemannian manifold M isometrically immersed in $\bar{M}$ is called an anti-invariant submanifold of $\bar{M}$ (or totally real submanifold of $\bar{M}$) if $JT_x(M) \subset T_x(M)^\perp$ for each point x of M, where $T_x(M)$ denotes the tangent space of M at x and $T_x(M)^\perp$ the normal space of M at x. If X is a tangent vector of M at x, then JX is a normal vector to M. Therefore we have $2m-n \geq n$, that is, $m \geq n$.

By a plane section we mean a 2-dimensional subspace of tangent space $T_x(M)$. A plane section p is said to be anti-invariant if Jp is perpendicular to p.

Proposition 5.1 (Chen-Ogiue [6]). Let M be a submanifold immersed in an almost Hermitian manifold $\bar{M}$. Then M is an anti-invariant submanifold of $\bar{M}$ if and only if every plane section of M is anti-invariant.

Proof. Let X be an arbitrary vector in $T_x(M)$. Let $e_1=X, e_2, \ldots, e_n$ be a basis of $T_x(M)$. We denote by p_{ij} the plane section spanned by e_i and e_j. Assume that every plane section is anti-invariant. Then $Jp_{1j} \cap p_{1j} = \{0\}$ for $j = 2, \ldots, n$. Thus JX is perpendicular to $e_1, \ldots, e_n$. Thus we have $JX \in T_x(M)^{\perp}$. This implies that M is anti-invariant in $\bar{M}$. The converse is clear.

In the sequel, as an ambient manifold, we take a complex space form $\bar{M}^m(c)$ of real dimension 2m and of constant holomorphic sectional curvature c. A submanifold M of $\bar{M}^m(c)$ is called an invariant submanifold with respect to the curvature transformation if each tangent space of M is invariant under the curvature transformation, that is, $K(X,Y)T_x(M) \subset T_x(M)$ for all vector fields X and Y on M, where K is the Riemannian curvature tensor of $\bar{M}^m(c)$.

Proposition 5.2 (Chen-Ogiue [6]). Let M be an n-dimensional submanifold of $\bar{M}^m(c)$ with $c \neq 0$. Then M is a Kaehlerian submanifold or an anti-invariant submanifold of $\bar{M}^m(c)$ if and only if M is an invariant submanifold with respect to the curvature transformation of $\bar{M}^m(c)$.

Proof. From (6.2) of Chapter I, we have

$$K(X,Y)Z = \frac{1}{4}c(g(Y,Z)X - g(X,Z)Y + \bar{g}(JY,Z)JX - \bar{g}(JX,Z)JY + 2\bar{g}(X,JY)JZ)$$

for any vector fields X, Y and Z on M. Thus if M is a Kaehlerian submanifold or an anti-invariant submanifold, then we easily see that $K(X,Y)Z \in T_x(M)$. Conversely, assume that $K(X,Y)Z \in T_x(M)$ for any X, Y and Z.

Putting $Z = X$ in the equation above, we have

$$K(X,Y)Z = \frac{1}{4}c(g(Y,X)X - g(X,X)Y + 3\bar{g}(X,JY)JX).$$

Since $K(X,Y)X \in T_x(M)$, we have $\bar{g}(X,JY)JX \in T_x(M)$ so that either $JX \in T_x(M)$ or $\bar{g}(X,JY) = 0$. This implies that $JT_x(M) \subset T_x(M)$ or $JT_x(M) \subset T_x(M)^{\perp}$, that is, M is invariant by J or anti-invariant by J.

Let CP^m be a complex projective space of constant holomorphic sectional curvature c (> 0). As usual we denote the homogeneous coordinates of CP^m by $z^0, z^1, \ldots, z^n$, where $z^h = x^h + iy^h$, $0 \leq h \leq m$. If M is an n-dimensional totally geodesic submanifold of CP^m, then the equation (2.3) shows that $K(X,Y)Z \in T_x(M)$ for any vectors X, Y and Z of $T_x(M)$. Thus Proposition 5.2 implies the following

Theorem 5.1 (Abe [1]). An n-dimensional complete totally geodesic submanifold M of CP^m is either a complex projective subspace or a real projective subspace. If M is a Kaehlerian submanifold, then M is given by the natural imbedding of C^{n+1} into C^{m+1}, i.e., $(z^0,\ldots,z^n) \longmapsto (z^0,\ldots,z^n,0,\ldots,0)$ in C^{m+1}. If M is a real projective space, by taking real and complex homogeneous coordinates properly in M and CP^m, respectively, the imbedding of M into CP^m is given as follows. Let $(x^0,\ldots,x^n)$ and $(z^0,\ldots,z^m)$ be the homogeneous coordinates of M and CP^m, respectively. Our imbedding is given by the natural imbedding of R^{n+1} into C^{m+1}, i.e., $(x^0,\ldots,x^n) \longmapsto (x^0,\ldots,x^n,0,\ldots,0)$. In this case, M is of constant curvature $\frac{1}{4}c$.

§6. Sasakian submanifolds

Let $\bar{M}$ be a (2m+1)-dimensional Sasakian manifold with structure tensors $(\phi,\xi,\eta,\bar{g})$. A submanifold N of $\bar{M}$ is called a Sasakian submanifold of $\bar{M}$ if the structure vector field ξ is tangent to M everywhere on M and ϕX is tangent to M for any tangent vector X to M. Obviously, M is a Sasakian manifold with respect to the induced structure tensors, which will be denoted by (ϕ,ξ,η,g).

Let $\bar{\nabla}$ (resp. ∇) be the operator of covariant differentiation with respect to $\bar{g}$ (resp. g). For any vector field X on M, we have

$$\bar{\nabla}_X\xi = \phi X, \qquad \bar{\nabla}_X\xi = \nabla_X\xi + B(X,\xi).$$

Since ϕX is tangent to M, we obtain $B(X,\xi) = 0$. Moreover, for any vector fields X and Y on M, we obtain

$$\bar{\nabla}_X\phi Y = \nabla_X\phi Y + B(X,\phi Y) = (\nabla_X\phi)Y + \phi\nabla_X Y + B(X,\phi Y)$$

$$= - g(X,Y)\xi + \eta(Y)X + \phi\nabla_X Y + B(X,\phi Y),$$

$$\bar{\nabla}_X\phi Y = (\bar{\nabla}_X\phi)Y + \phi\bar{\nabla}_X Y = - g(X,Y)\xi + \eta(Y)X + \phi\nabla_X Y + \phi B(X,Y).$$

Consequently we have the following equations:

$$(6.1) \qquad B(X,\phi Y) = B(\phi X,Y) = \phi B(X,Y), \quad B(X,\xi) = 0.$$

<u>Proposition 6.1 (Tanno [60], Yano-Ishihara [75])</u>. Any Sasakian submanifold M is a minimal submanifold.

<u>Proof</u>. For each point x of M, we can take a basis $\xi,e_1,\ldots,e_n,\phi e_1,\ldots,\phi e_n$ in $T_x(M)$. By (6.1) we see that

$$\sum_{i=1}^{n} (B(e_i,e_i) + B(\phi e_i,\phi e_i)) + B(\xi,\xi) = 0$$

because of

$$B(\phi e_i,\phi e_i) = \phi^2 B(e_i,e_i) = - B(e_i,e_i) + \eta(B(e_i,e_i))\xi = - B(e_i,e_i).$$

This means that M is a minimal submanifold.

From the Gauss equation (2.4) and (6.1) we obtain

Proposition 6.2. Let M be a Sasakian submanifold of a Sasakian manifold $\bar{M}$. Then the respective curvature tensors R and K of M and $\bar{M}$ satisfy

$$R(X,\phi X,X,\phi X) = K(X,\phi X,X,\phi X) - 2\bar{g}(B(X,X),B(X,X)).$$

From Proposition 6.2 we have the following

Proposition 6.3. Let M be a Sasakian submanifold of a Sasakian space form $\bar{M}(c)$ with constant ϕ-sectional curvature c. Then M is totally geodesic if and only if M is of constant ϕ-sectional curvature c.

Proposition 6.4 (Kon [22]). If the second fundamental form B of a Sasakian submanifold M is parallel, then M is totally geodesic.

Proof. From (2.1) and (6.1) we see that

$$(\nabla_X B)(Y,\xi) = - B(Y,\phi X) = - \phi B(X,Y)$$

for any vector fields X and Y on M. Thus we have

$$\phi^2 B(X,Y) = - B(X,Y) + \bar{g}(\xi,B(X,Y))\xi = - B(X,Y) = 0,$$

which shows that M is totally geodesic.

Example 6.1. Let S^{2m+1} be a unit sphere with standard Sasakian structure as in Example in §7 of Chapter I. An odd-dimensional unit sphere S^{2n+1} $(n < m)$ with induced structure is a totally geodesic Sasakian submanifold of S^{2m+1}. Obviously the Sasakian space form $E^{2n+1}(-3)$ in $E^{2m+1}(-3)$ is a totally geodesic Sasakian submanifold.

Example 6.2. The circle bundle (Q^n,S^1) over a hyperquadric in CP^{n+1} is a Sasakian submanifold of S^{2n+3} which is an η-Einstein manifold, i.e., the Ricci tensor S of (Q^n,S^1) is of the form $S(X,Y) = ag(X,Y) + b\eta(X)\eta(Y)$, for some constants a and b.

§7. Anti-invariant submanifolds of contact manifolds

Let $\bar{M}$ be a (2m+1)-dimensional almost contact metric manifold with structure tensors $(\phi,\xi,\eta,\bar{g})$. An n-dimensional submanifold immersed in $\bar{M}$ is said to be anti-invariant in $\bar{M}$ if $\phi T_x(M) \subset T_x(M)^{\perp}$ for each $x \in M$. Then we see that, ϕ being of rank 2m, $n \leq m+1$. When $n = m+1$, we have the following

Proposition 7.1 (Yano-Kon [79]). Let $\bar{M}$ be an almost contact metric manifold of dimension 2n+1 and let M be an anti-invariant submanifold of $\bar{M}$ of dimension n+1. Then the structure vector field ξ is tangent to M.

Proof. By the assumption we have $\phi T_x(M) = T_x(M)^{\perp}$ at each point x of M. For any tangent vector field X on M we have

$$\bar{g}(\xi,\phi X) = -\bar{g}(\phi\xi,X) = 0.$$

Consequently, the structure vector field ξ is tangent to M.

In the following, we denote by $\bar{\nabla}$ the operator of covariant differentiation on $\bar{M}$ and ∇ that on M.

Proposition 7.2. Let M be a submanifold of a Sasakian manifold $\bar{M}$ tangent to the structure vector field ξ of $\bar{M}$. Then ξ is parallel with respect to the induced connection on M, i.e., $\nabla_X\xi = 0$ for any vector field X on M if and only if M is an anti-invariant submanifold in $\bar{M}$.

Proof. Since ξ is tangent to M, we have

$$\phi X = \nabla_X\xi + B(X,\xi).$$

Thus if ξ is parallel, then $\phi X = B(X,\xi)$ is normal to M. Conversely if ϕX is normal to M, then ξ is parallel.

Proposition 7.3. Let M be an n-dimensional submanifold of a (2m+1)-dimensional Sasakian manifold $\bar{M}$. If the structure vector field ξ is normal to M, then M is an anti-invariant submanifold in $\bar{M}$ and $n \leq m$.

Proof. Since ξ is normal to M, the Weingarten formula implies

$$\bar{g}(\phi X,Y) = \bar{g}(\bar{\nabla}_X\xi,Y) = g(-A_\xi X,Y) + \bar{g}(D_X\xi,Y) = - g(A_\xi X,Y)$$

for any vector fields X and Y on M. Since A_ξ is symmetric and ϕ is anti-symmetric, we have $A_\xi = 0$ and ϕX is normal to M. Thus M is anti-invariant in $\bar{M}$ and $n \leq m$.

In view of Propositions 7.1, 7.2 and 7.3, we have only to consider anti-invariant submanifolds tangent to the structure vector field and those normal to the structure vector field.

In Chapter IV we study anti-invariant submanifolds of Sasakian manifolds tangent to the structure vector field and in Chapter V we study anti-invariant submanifolds of Sasakian manifolds normal to the structure vector field.

Chapter III

Anti-invariant submanifolds of Kaehlerian manifolds

In this chapter we shall study anti-invariant submanifolds of Kaehlerian manifolds. Throughout the chapter, we denote by $\bar{M}$ the ambient Kaehlerian manifold with the almost complex structure J and the Hermitian metric $\bar{g}$ and by M an anti-invariant submanifold immersed in $\bar{M}$ with the induced metric g. (For anti-invariant submanifolds immersed in complex manifolds, see Lai [33], Wells [62]).

§1. Local formulas and basic properties

Let $\bar{M}$ be a real $2m$-dimensional Kaehlerian manifold and let M be an n-dimensional anti-invariant submanifold immersed in $\bar{M}$. We choose a local field of orthonormal frames $e_1,\dots,e_n;e_{n+1},\dots,e_m;e_{1^*}=Je_1,\dots,e_{n^*}=Je_n;$ $e_{(n+1)^*}=Je_{n+1},\dots,e_{m^*}=Je_m$ in $\bar{M}$ in such a way that, restricted to M, $e_1,\dots,$ e_n are tangent to M and hence all the remaining vectors are normal to M. With respect to this frame field of $\bar{M}$, let $\omega^1,\dots,\omega^n;\omega^{n+1},\dots,\omega^m;\omega^{1^*},\dots,$ $\omega^{n^*};\omega^{(n+1)^*},\dots,\omega^{m^*}$ be the field of dual frames. Unless otherwise stated, we use the conventions that the ranges of indices are respectively:

$$A, B, C, D = 1,\dots,m,1^*,\dots,m^*,$$

$$i, j, k, l, t, s = 1,\dots,n,$$

$$a, b, c, d = n+1,\dots,m,1^*,\dots,m^*,$$

$$\alpha, \beta, \gamma = n+1,\dots,m,$$

$$\lambda, \mu, \nu = n+1,\dots,m,(n+1)^*,\dots,m^*.$$

In view of equations (6.3) of Chapter I, it follows that the connection forms (ω_B^A) of $\bar{M}$ satisfy the following equations:

$$\omega_j^i + \omega_i^j = 0, \qquad \omega_j^i = \omega_{j*}^{i*}, \qquad \omega_j^{i*} = \omega_i^{j*},$$

(1.1)
$$\omega_\beta^\alpha + \omega_\alpha^\beta = 0, \qquad \omega_\beta^\alpha = \omega_{\beta*}^{\alpha*}, \qquad \omega_\beta^{\alpha*} = \omega_\alpha^{\beta*},$$

$$\omega_\alpha^i + \omega_i^\alpha = 0, \qquad \omega_\alpha^i = \omega_{\alpha*}^{i*}, \qquad \omega_\alpha^{i*} = \omega_i^{\alpha*}.$$

From (3.4) of Chapter II and (1.1), we see that the second fundamental form h_{ij}^a of M satisfies

(1.2)
$$h_{jk}^i = h_{ik}^j = h_{ij}^k,$$

where we write h_{jk}^{i*} as h_{jk}^i. That is, the second fundamental form A of M satisfies

$$A_{JX}Y = A_{JY}X$$

for any vector fields X and Y on M.

In the sequel, we assume that the ambient manifold $\bar{M}$ is of constant holomorphic sectional curvature c and of real dimension 2m and denote it by $\bar{M}^m(c)$. The curvature tensor K of $\bar{M}^m(c)$ is given by

(1.3)
$$K_{BCD}^A = \frac{1}{4}c(\delta_{AC}\delta_{BD} - \delta_{AD}\delta_{BC} + J_{AC}J_{BD} - J_{AD}J_{BC} + 2J_{AB}J_{CD}).$$

Since M is anti-invariant in $\bar{M}^m(c)$, we have

(1.4)
$$K_{jkl}^i = \frac{1}{4}c(\delta_{ik}\delta_{jl} - \delta_{il}\delta_{jk}).$$

Then the Gauss equation (3.7) of Chapter II implies

(1.5) $$R^{i}_{jkl} = \frac{1}{4}c(\delta_{ik}\delta_{jl} - \delta_{il}\delta_{jk}) + \sum_{a}(h^{a}_{ik}h^{a}_{jl} - h^{a}_{il}h^{a}_{jk}).$$

Therefore we have

Proposition 1.1. Let M be an anti-invariant submanifold of a complex space form $\bar{M}^m(c)$. If M is totally geodesic, then M is of constant curvature $\frac{1}{4}c$.

From (1.5), we see that the Ricci tensor R_{ij} of M is given by

(1.6) $$R_{ij} = \frac{1}{4}c(n-1)\delta_{ij} + \sum_{a,k}(h^{a}_{kk}h^{a}_{ij} - h^{a}_{ik}h^{a}_{jk}).$$

Therefore the scalar curvature r of M is given by

(1.7) $$r = \frac{1}{4}n(n-1)c + \sum_{a,i,j}(h^{a}_{ii}h^{a}_{jj} - h^{a}_{ij}h^{a}_{ij}).$$

If M is minimal in $\bar{M}(c)$, i.e., $\sum_{i} h^{a}_{ii} = 0$, then (1.6), (1.7) can be rewritten respectively as

(1.8) $$R_{ij} = \frac{1}{4}(n-1)c\delta_{ij} - \sum_{a,k} h^{a}_{ik}h^{a}_{jk},$$

(1.9) $$r = \frac{1}{4}n(n-1)c - \sum_{a,i,j}(h^{a}_{ij})^2.$$

From these equations we obtain the following

Proposition 1.2. Let M be an n-dimensional anti-invariant minimal submanifold of a complex space form $\bar{M}^m(c)$ of real dimension 2m. Then for the Ricci tensor S and the scalar curvature r of M we have

(a) $S - \frac{1}{4}(n-1)cg$ is negative semi-definite,

(b) $r \leq \frac{1}{4}n(n-1)c$.

Proposition 1.3. Let M be an n-dimensional anti-invariant minimal submanifold of a complex space form $\bar{M}^m(c)$. Then M is totally geodesic if and only if M satisfies one of the following conditions:

(a) M is of constant curvature $\frac{1}{4}c$,

(b) $S = \frac{1}{4}(n-1)cg$,

(c) $r = \frac{1}{4}n(n-1)c$.

Propositions 1.1, 1.2 and 1.3 are due to Chen-Ogiue [6].

§2. f-structure in the normal bundle

Let M be an n-dimensional anti-invariant submanifold of a real 2m-dimensional Kaehlerian manifold $\bar{M}$. Then we see that $JT_x(M) \subset T_x(M)^\perp$ at each point x of M. Let $N_x(M)$ be an orthogonal complement of $JT_x(M)$ in $T_x(M)^\perp$. Then we have the decomposition:

$$T_x(M)^\perp = JT_x(M) \oplus N_x(M).$$

Thus we see that the space $N_x(M)$ is invariant under the action of J, that is, if $N \in N_x(M)$, then $JN \in N_x(M)$. If N is a vector field in the normal bundle $T(M)^\perp$, we put

$$JN = PN + fN, \tag{2.1}$$

where PN is the tangential part of JN and fN the normal part of JN. Then P is a tangent bundle valued 1-form on the normal bundle and f is an endomorphism of the normal bundle. Then, putting N = JX in (2.1) for tangent vector field X in M and applying J to (2.1), we find

$$\begin{aligned} &PfN = 0, \qquad f^2N = -N - JPN, \\ &PJX = -X, \qquad fJX = 0, \end{aligned} \tag{2.2}$$

where X is a tangent vector field in M and N is a vector field in the normal bundle. Equations (2.2) imply that

$$f^3 + f = 0.$$

Therefore, f being of constant rank, if f does not vanish, then it defines an f-structure in the normal bundle. (For the f-structure, see Yano [66].) From (2.1), using the Gauss and Weingarten formulas, we have

$$- JA_N X + fD_X N = B(X,PN) + D_X(fN), \tag{2.3}$$

from which

$$(D_X f)N = - B(X,PN) - JA_N X, \tag{2.4}$$

where $(D_X f)N = D_X(fN) - f(D_X N)$. If $D_X f = 0$ for all tangent vector field X, then the f-structure in the normal bundle is said to be parallel.

Lemma 2.1. Let M be an n-dimensional anti-invariant submanifold of a real 2m-dimensional Kaehlerian manifold $\bar{M}$. If the f-structure in the normal bundle is parallel, then we have

$$A_N = 0 \quad \text{for} \quad N \in N_x(M). \tag{2.5}$$

Proof. If $N \in N_x(M)$, then we have $PN = 0$. Thus by the assumption and (2.4) we have $JA_N X = 0$ and hence $A_N X = 0$ for all X, which proves (2.5).

Remark. We can take a basis $e_{1^*}, \ldots, e_{n^*}$ for $JT_x(M)$ and a basis $e_{n+1}, \ldots, e_m, e_{(n+1)^*}, \ldots, e_{m^*}$ for $N_x(M)$. Therefore if the f-structure in the normal bundle is parallel, then we have

$$A_\lambda = 0, \qquad \text{i.e.,} \qquad h^\lambda_{ij} = 0. \tag{2.6}$$

We next assume that M is an n-dimensional anti-invariant submanifold of a complex space form $\bar{M}^m(c)$ with parallel f-structure in the normal bundle. Then equations (1.5), (1.6) and (1.7) can be rewritten as follows:

$$R^i_{jkl} = \frac{1}{4}c(\delta_{ik}\delta_{jl} - \delta_{il}\delta_{jk}) + \sum_t (h^t_{ik}h^t_{jl} - h^t_{il}h^t_{jk}), \tag{2.7}$$

$$R_{ij} = \frac{1}{4}(n-1)c\delta_{ij} + \sum_{t,k} (h^t_{kk}h^t_{ij} - h^t_{ik}h^t_{jk}), \tag{2.8}$$

$$r = \frac{1}{4}n(n-1)c + \sum_{t,i,j} (h^t_{ii}h^t_{jj} - h^t_{ij}h^t_{ij}). \tag{2.9}$$

Moreover, if M is minimal, we have

$$R_{ij} = \frac{1}{4}(n-1)c\delta_{ij} - \sum_{t,k} h^t_{ik}h^t_{jk}, \tag{2.10}$$

$$r = \frac{1}{4}n(n-1)c - \sum_{t,i,j} (h^t_{ij})^2. \tag{2.11}$$

§3. Integral formulas

Let $\bar{M}^m(c)$ be a complex space form of real dimension 2m and of constant holomorphic sectional curvature c and let M be an anti-invariant submanifold of dimension n of $\bar{M}^m(c)$. Then, by (6.2) of Chapter II and the Ricci equation (2.5) of Chapter II, we see that the second fundamental form of M satisfies the Codazzi equation (2.7) of Chapter II, i.e., $h^a_{ijk} - h^a_{ikj} = 0$. Therefore, substituting (1.3) into (3.18) of Chapter II, we obtain

Lemma 3.1. Let M be an n-dimensional anti-invariant submanifold of a complex space form $\bar{M}^m(c)$. Then we have

$$\begin{aligned}\sum_{a,i,j} h^a_{ij}\Delta h^a_{ij} = & \sum_{a,i,j,k} h^a_{ij}h^a_{kkij} + \sum_a [\tfrac{1}{4}nc\mathrm{Tr}A_a^2 - \tfrac{1}{4}c(\mathrm{Tr}A_a)^2] \\ & + \sum_t [\tfrac{1}{4}c\mathrm{Tr}A_t^2 - \tfrac{1}{4}c(\mathrm{Tr}A_t)^2] + \sum_{a,b} \{\mathrm{Tr}(A_aA_b - A_bA_a)^2 \\ & - [\mathrm{Tr}(A_aA_b)]^2 - \mathrm{Tr}A_b\mathrm{Tr}(A_aA_bA_a)\},\end{aligned} \tag{3.1}$$

where we have put $A_t = A_{t*}$.

Using Lemma 2.1 and (3.1), we obtain the following lemmas:

Lemma 3.2. Let M be an n-dimensional anti-invariant submanifold of a complex space form $\bar{M}^m(c)$. If the f-structure in the normal bundle is parallel, then we have

$$\sum_{a,i,j} h_{ij}^a \Delta h_{ij}^a = \sum_{a,i,j,k} h_{ij}^a h_{kkij}^a + \sum_t [\tfrac{1}{4}(n+1)c\,\mathrm{Tr}A_t^2 - \tfrac{1}{2}c(\mathrm{Tr}A_t)^2] \tag{3.2}$$

$$+ \sum_{t,s} \{\mathrm{Tr}(A_tA_s - A_sA_t)^2 - [\mathrm{Tr}(A_tA_s)]^2 + \mathrm{Tr}A_s\mathrm{Tr}(A_tA_sA_t)\}.$$

Lemma 3.3. Let M be an n-dimensional anti-invariant minimal submanifold of a complex space form $\bar{M}^m(c)$. Then we have

$$\sum_{a,i,j} h_{ij}^a \Delta h_{ij}^a = \tfrac{1}{4}nc \sum_a \mathrm{Tr}A_a^2 + \tfrac{1}{4}c\sum_t \mathrm{Tr}A_t^2 \tag{3.3}$$

$$+ \sum_{a,b} \{\mathrm{Tr}(A_aA_b - A_bA_a)^2 - [\mathrm{Tr}(A_aA_b)]^2\}.$$

Lemma 3.4. Let M be an n-dimensional anti-invariant minimal submanifold of a complex space form $\bar{M}^m(c)$. If the f-structure in the normal bundle is parallel, then we have

$$\sum_{a,i,j} h_{ij}^a \Delta h_{ij}^a = \tfrac{1}{4}(n+1)c\sum_t \mathrm{Tr}A_t^2 + \sum_{t,s} \{\mathrm{Tr}(A_tA_s - A_sA_t)^2 \tag{3.4}$$

$$- [\mathrm{Tr}(A_tA_s)]^2\}.$$

In the sequel, we need the following lemma proved in Chern-do Carmo-Kobayashi [10].

Lemma 3.5 ([10]). Let A and B be symmetric (n,n)-matrices. Then

$$-\operatorname{Tr}(AB - BA)^2 \leq 2\operatorname{Tr}A^2\operatorname{Tr}B^2$$

and the equality holds for non-zero matrices A and B if and only if A and B can be transformed by an orthogonal matrix simultaneousely into scalar multiples of $\bar{A}$ and $\bar{B}$ respectively, where

$$\bar{A} = \left(\begin{array}{cc|c} 0 & 1 & \\ & & 0 \\ 1 & 0 & \\ \hline \multicolumn{2}{c|}{0} & 0 \end{array}\right), \qquad \bar{B} = \left(\begin{array}{cc|c} 1 & 0 & \\ & & 0 \\ 0 & -1 & \\ \hline \multicolumn{2}{c|}{0} & 0 \end{array}\right).$$

Moreover, if A_1, A_2, A_3 are symmetric (n,n)-matrices such that

$$-\operatorname{Tr}(A_aA_b - A_bA_a)^2 = 2\operatorname{Tr}A_a^2\operatorname{Tr}A_b^2, \quad 1 \leq a,b \leq 3, \quad a \neq b,$$

then at least one of the matrices A_a must be zero.

We now put

$$S_{ab} = \sum_{i,j} h_{ij}^a h_{ij}^b = \operatorname{Tr}A_aA_b, \quad S_a = S_{aa}, \quad S = \sum_a S_a,$$

so that S_{ab} is a symmetric (n,n)-matrix and can be assumed to be diagonal for a suitable frame. S is the square of the length of the second fundamental form of M. When the f-structure in the normal bundle is parallel, using these notations, we can rewrite (3.2) in the following form:

$$(3.5) \quad \sum_{a,i,j} h_{ij}^a \Delta h_{ij}^a = \sum_{a,i,j,k} h_{ij}^a h_{kkij}^a + \tfrac{1}{4}(n+1)cS - \sum_t S_t^2$$
$$+ \sum_{t,s} \operatorname{Tr}(A_tA_s - A_sA_t)^2 - \tfrac{1}{2}c\sum_t(\operatorname{Tr}A_t)^2 + \sum_{t,s} \operatorname{Tr}A_s \operatorname{Tr}(A_tA_sA_t).$$

On the other hand, using Lemma 3.5, we have the following inequality which plays an important rôle in the sequel.

$$(3.6)\qquad -\sum_{t,s}\mathrm{Tr}(A_tA_s - A_sA_t)^2 + \sum_t S_t^2 - \frac{1}{4}(n+1)cS$$

$$\le 2\sum_{t\neq s} S_tS_s + \sum_t S_t^2 - \frac{1}{4}(n+1)cS$$

$$= [(2 - \frac{1}{n})S - \frac{1}{4}(n+1)c]S - \frac{1}{n}\sum_{t>s}(S_t - S_s)^2.$$

Now we assume that M is minimal. Then (3.5) and (3.6) imply

$$(3.7)\qquad -\sum_{a,i,j} h_{ij}^a \Delta h_{ij}^a \le [(2 - \frac{1}{n})S - \frac{1}{4}(n+1)c]S - \frac{1}{n}\sum_{t>s}(S_t - S_s)^2.$$

If M is compact and orientable, we see that

$$\int_M \sum_{a,i,j,k}(h_{ijk}^a)^2 *1 = -\int_M \sum_{a,i,j} h_{ij}^a \Delta h_{ij}^a *1.$$

Consequently, we have the following

Theorem 3.1. Let M be an n-dimensional compact orientable anti-invariant minimal submanifold of a complex space form $\bar{M}^m(c)$. If the f-structure in the normal bundle is parallel, then we have

$$(3.8)\qquad \int_M [(2 - \frac{1}{n})S - \frac{1}{4}(n+1)c]S*1 \ge \int_M \sum_{a,i,j,k}(h_{ijk}^a)^2 *1 \ge 0.$$

Corollary 3.1. Let M be an n-dimensional compact orientable anti-invariant minimal submanifold of a complex space form $\bar{M}^m(c)$ with parallel f-structure in the normal bundle. If $S < n(n+1)c/4(2n-1)$, then M is totally geodesic.

When $n = m$, we always have $f = 0$. Thus we obtain the following results proved in Chen-Ogiue [6].

Theorem 3.2. Let M be an n-dimensional compact orientable anti-invariant submanifold of a complex space form $\bar{M}^n(c)$. If M is minimal, then we have

$$(3.9)\qquad \int_M [(2-\tfrac{1}{n})S - \tfrac{1}{4}(n+1)c]S{*}1 \geq \int_M \sum_{t,i,j,k}(h^t_{ijk})^2{*}1 \geq 0.$$

Corollary 3.2. Let M be an n-dimensional compact orientable anti-invariant minimal submanifold of a complex space form $\bar{M}^n(c)$. If $S < n(n+1)c/4(2n-1)$, then M is totally geodesic.

Without the assumption on the f-structure in the normal bundle, Ludden-Okumura-Yano [37] proved

Theorem 3.3. Let M be an n-dimensional compact orientable anti-invariant minimal submanifold of a complex space form $\bar{M}^m(c)$ $(c > 0)$. If $S \leq \frac{1}{4}ncq/(2q-1)$, where $q = 2m-n$, then M is totally geodesic.

Proof. From (3.3) and Lemma 3.5, we find

$$(3.10)\qquad \sum_{a,i,j,k}(h^a_{ijk})^2 - \tfrac{1}{2}\Delta S = -\sum_{a,b}\mathrm{Tr}(A_aA_b - A_bA_a)^2 + \sum_a S_a^2 - \tfrac{1}{4}ncS - \tfrac{1}{4}c\sum_t \mathrm{Tr}A_t^2$$
$$\leq [(2-\tfrac{1}{q})S - \tfrac{1}{4}nc]S - \tfrac{1}{q}\sum_{a>b}(S_a - S_b)^2 - \tfrac{1}{4}c\sum_t \mathrm{Tr}A_t^2.$$

Since M is compact and orientable, we have

$$\int_M \sum_{a,i,j,k}(h^a_{ijk})^2{*}1 \leq \int_M [(2-\tfrac{1}{q})S - \tfrac{1}{4}nc]S{*}1.$$

Therefore, by the assumption and Lemma 3.5, we see that

$$(3.11)\qquad \sum_{a>b}(S_a - S_b)^2 = 0, \qquad \sum_t \mathrm{Tr}A_t^2 = 0.$$

Equations (3.11) show that $A_t = 0$ for all t and hence $S_t = 0$ and $S_a = S_b$ for all a and b. Thus we have $S_a = 0$ and hence $A_a = 0$ for all a.

§4. Anti-invariant minimal submanifolds

In this section we study n-dimensional anti-invariant minimal submanifolds with parallel f-structure in the normal bundle of a complex space form under the assumption that $S = n(n+1)c/4(2n-1)$. Thus if the submanifold is not totally geodesic, we have $c > 0$. In the sequel, we assume that $c = 4$ without loss of generality. First of all, we prove the following

Theorem 4.1. Let M be an n-dimensional $(n > 1)$ anti-invariant minimal submanifold with parallel f-structure in the normal bundle of a complex space form $\bar{M}^m(4)$. If $S = n(n+1)/(2n-1)$, then $n = 2$ and M is a flat surface of some $\bar{M}^2(4)$ in $\bar{M}^m(4)$, where $\bar{M}^2(4)$ is a totally geodesic Kaehlerian submanifold of real dimension 4 of $\bar{M}^m(4)$. With respect to an adapted dual orthonormal frame field $\omega^1, \omega^2, \omega^{1*}, \omega^{2*}$ in $\bar{M}^2(4)$, the connection form (ω^A_B) of $\bar{M}^2(4)$, restricted to M, is given by

$$\begin{pmatrix} 0 & 0 & -\lambda\omega^2 & -\lambda\omega^1 \\ 0 & 0 & -\lambda\omega^1 & \lambda\omega^2 \\ \lambda\omega^2 & \lambda\omega^1 & 0 & 0 \\ \lambda\omega^1 & -\lambda\omega^2 & 0 & 0 \end{pmatrix}, \qquad \lambda = \frac{1}{\sqrt{2}}.$$

Proof. From the assumption and (3.8), the second fundamental form of M is parallel, i.e., $h^a_{ijk} = 0$. Thus (3.5), (3.6) and (3.7) together with Lemma 3.5, imply

$$\sum_{t,s} (S_t - S_s)^2 = 0, \tag{4.1}$$

$$-\mathrm{Tr}(A_tA_s - A_sA_t)^2 = 2\mathrm{Tr}A_t^2\mathrm{Tr}A_s^2. \tag{4.2}$$

In view of Lemma 3.5 we may assume that $A_t = 0$ for $t = 3,\dots,n$, which means that $S_t = 0$ for $t = 3,\dots,n$. On the other hand, by (4.1) we have $S_t = S_s$ for all t and s. Therefore we must have $n = 2$. Thus Lemma 3.5 implies that

$$(4.3) \qquad A_1 = \lambda \begin{pmatrix} 0 & 1 \\ 1 & 0 \end{pmatrix}, \qquad A_2 = \lambda \begin{pmatrix} 1 & 0 \\ 0 & -1 \end{pmatrix},$$

because of $h^1_{12} = \lambda = h^2_{11}$. Since the f-structure in the normal bundle is parallel, we have $h^{\lambda}_{ij} = 0$ by Lemma 2.1. Thus we have $\omega^{\lambda}_i = h^{\lambda}_{ij}\omega^j = 0$, that is,

$$(4.4) \qquad \omega^{\lambda}_i = 0.$$

From (3.10) of Chapter II we also have

$$(4.5) \qquad dh^a_{ij} = h^a_{i1}\omega^1_i + h^a_{1j}\omega^1_i - h^b_{ij}\omega^a_b.$$

Putting $a = \lambda$ in (4.5), we have $h^k_{ij}\omega^{\lambda}_{k*} = 0$, $k = 1,2$. From this and (4.3) we see that

$$(4.6) \qquad \omega^{\lambda}_{i*} = 0.$$

Putting $a = 1$, $i = 1$ and $j = 2$ in (4.5), we deduce that $d\lambda = dh^1_{12} = 0$, which means that λ is constant. Since $S = 2$, we have $2\lambda^2 = 1$. Thus we may assume that $\lambda = 1/\sqrt{2}$. Moreover, since M is not totally geodesic, i.e., $\lambda \neq 0$, we have

$$(4.7) \qquad \omega^{t*}_i \neq 0, \qquad t = 1,2.$$

Since we have (4.4). (4.6) and (4.7) we can consider a distribution L defined by

$$\omega^{\lambda} = 0, \qquad \omega^{\lambda}_{i} = 0, \qquad \omega^{\lambda}_{i*} = 0.$$

It easily follows from the structure equations that

$$d\omega^{\lambda} = 0, \qquad d\omega^{\lambda}_{i} = 0, \qquad d\omega^{\lambda}_{i*} = 0.$$

Therefore the distribution L is a real 4-dimensional completely integrable distribution. We consider the maximal integral submanifold $\bar{M}(x)$ of L through $x \in M$. Then $\bar{M}(x)$ is of real dimension 4 and by construction it is totally geodesic Kaehlerian submanifold in $\bar{M}^m(4)$ and hence is of constant holomorphic sectional curvature 4. Moreover M is anti-invariant in $\bar{M}^2(4)$. In $\bar{M}^2(4)$ in $\bar{M}^m(4)$, (4.3) and (4.5) imply the equations:

$$\text{(4.8)} \qquad \begin{aligned} &\omega^{1*}_{1} = \lambda\omega^{2}, \qquad && \omega^{1*}_{2} = \omega^{2*}_{1} = \lambda\omega^{1}; \\ &\omega^{2*}_{2} = -\lambda\omega^{2}, && \omega^{2}_{1} = \omega^{2*}_{1*} = 0. \end{aligned}$$

From (1.5) and (4.3) we have

$$R^{1}_{212} = 1 - \frac{1}{2} - \frac{1}{2} = 0.$$

Consequently, M is a flat surface. From these considerations we have our assertion.

<u>Example 4.1</u>. Let S^5 be a 5-dimensional unit sphere with standard Sasakian structure $(\phi, \xi, \eta, \bar{g})$. The integral curves of the structure vector field ξ are great circles S^1 in S^5 which are the fibres of the standard fibration $\pi : S^5 \longrightarrow CP^2$ onto complex projective space CP^2 of real dimension 4 and of constant holomorphic sectional curvature 4. Let $T = S^1 \times S^1$ be a maximal torus which is imbedded in S^5 as an anti-invariant submanifold

normal to the structure vector field ξ. The imbedding $X : T \longrightarrow S^5$ is given by

$$X = \frac{1}{\sqrt{3}}(\cos u^1, \sin u^1, \cos u^2, \sin u^2, \cos u^3, \sin u^3),$$

where $u^3 = -u^1 - u^2$ in C^3. Now we consider the following diagram:

$$\begin{array}{ccc} T & \xrightarrow{\ X\ } & S^5 \\ & \searrow & \downarrow \pi \\ & & CP^2 \end{array} .$$

We easily see that $\pi|X(T)$ is one to one. Consequently T is imbedded in CP^2 by $\pi \cdot X$. Thus $S = 2$. (For the detail, see Chapter V and also Ludden-Okumura-Yano [36].)

From this example and Theorem 4.1 we obtain

Theorem 4.2. Let M be an n-dimensional $(n > 1)$ compact orientable anti-invariant minimal submanifold of CP^m with parallel f-structure in the normal bundle. If $S = n(n+1)/(2n-1)$, then M is a torus $S^1 \times S^1$ in some CP^2 in CP^m.

In the case where $n = m$, we have

Theorem 4.3 ([36]). Let M be an n-dimensional $(n > 1)$ compact orientable anti-invariant minimal submanifold of CP^n. If $S = n(n+1)/(2n-1)$, then M is $S^1 \times S^1$ in CP^2.

§5. Commutative second fundamental forms

We first prepare some lemmas.

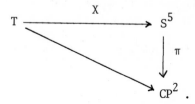

Lemma 5.1. Let M be an n-dimensional anti-invariant submanifold of a real 2m-dimensional Kaehlerian manifold $\bar{M}$. If the f-structure in the normal bundle is parallel, then M is flat if and only if the normal connection of M is flat, i.e., $R^a_{bkl} = 0$.

Proof. From Lemma 2.1, we get $h^\lambda_{ij} = 0$, which shows that $\omega^\lambda_i = 0$. Thus (3.6), (3.8) of Chapter II and (1.1) imply

$$\Omega^{i*}_{j*} = d\omega^{i*}_{j*} + \omega^{i*}_a \wedge \omega^a_{j*} = d\omega^i_j + \omega^i_k \wedge \omega^k_j = \Omega^i_j,$$

$$\Omega^{i*}_\lambda = 0, \qquad \Omega^\lambda_\mu = 0,$$

which show that $R^{i*}_{j*kl} = R^i_{jkl}$ and $R^\lambda_{\mu kl} = R^\lambda_{j*kl} = R^{i*}_{\mu kl} = 0$. Thus we have our assertion.

Lemma 5.2. Let M be an n-dimensional anti-invariant submanifold of a real 2m-dimensional Kaehlerian manifold $\bar{M}$. If the second fundamental forms of M are commutative, i.e., $A_aA_b = A_bA_a$ for all a and b, then we can choose an orthonormal frame for which A_t is of the form

$$A_t = \begin{pmatrix} 0 & & & & & \\ & \ddots & & & & \\ & & 0 & & & \\ & & & \lambda_t & & \\ & & & & 0 & \\ & & & & & \ddots \\ & & & & & & 0 \end{pmatrix} t, \qquad t = 1,\dots,n,$$

that is, $h^t_{ij} = 0$ unless $t = i = j$.

Proof. If $A_aA_b = A_bA_a$, we can choose an orthonormal frame $e_1,\dots,e_n$ for which all A_a are simultaneously diagonal, i.e., $h^a_{ij} = 0$ when $i \neq j$, that is, $h^t_{ij} = 0$ when $i \neq j$. Thus (1.2) shows that $h^t_{ij} = 0$ unless $t = i = j$.

Corollary 5.1. Let M be an n-dimensional anti-invariant minimal submanifold with commutative second fundamental forms of a real 2m-dimensional Kaehlerian manifold $\bar{M}$. If the f-structure in the normal bundle is parallel, then M is totally geodesic.

Proof. From Lemma 5.2, we have $\lambda_t = 0$ for all t, by the fact that $TrA_t = 0$. On the other hand, we know already that $A_\lambda = 0$. Thus M is totally geodesic.

Corollary 5.2. Let M be an n-dimensional $(n > 1)$ anti-invariant, totally umbilical submanifold of a real 2m-dimensional Kaehlerian manifold $\bar{M}$. If the f-structure in the normal bundle is parallel, then M is totally geodesic.

Proof. Since M is umbilical, we have $h_{ij}^t = \delta_{ij}(TrA_t)/n$ and we also have $A_\lambda = 0$ by Lemma 2.1. Therefore the second fundamental forms of M are commutative. Thus Lemma 5.2 implies that $h_{ij}^t = 0$ unless $t = i = j$. On the other hand, we have $h_{ij}^t = \lambda_t \delta_{ij}/n$. Putting $t \neq i = j$, we find $\lambda_t = 0$ and hence M is totally geodesic.

From Proposition 1.1 and Corollaries 5.1, 5.2 we have

Corollary 5.3. Under the same assumptions as in Corollary 5.1 or Corollary 5.2, if the ambient manifold $\bar{M}$ is of constant holomorphic sectional curvature c, then M is of constant curvature $\frac{1}{4}c$.

Lemma 5.3. Let M be an n-dimensional anti-invariant submanifold of a complex space form $\bar{M}^m(c)$ with parallel f-structure in the normal bundle. Then M is of constant curvature $\frac{1}{4}c$ if and only if the second fundamental forms of M are commutative.

Proof. First of all, we have $h_{ij}^\lambda = 0$ by Lemma 2.1 and hence (1.2) and (2.7) imply

$$R^i_{jkl} = \frac{1}{4}c(\delta_{ik}\delta_{jl} - \delta_{il}\delta_{jk}) + \sum_t (h^t_{ik}h^t_{jl} - h^t_{il}h^t_{jk})$$

$$= \frac{1}{4}c(\delta_{ik}\delta_{jl} - \delta_{il}\delta_{jk}) + \sum_t (h^i_{tk}h^j_{tl} - h^i_{tl}h^j_{tk}),$$

which proves our assertion.

Lemma 5.4. Let M be an n-dimensional anti-invariant submanifold of a real 2m-dimensional Kaehlerian manifold $\bar{M}$. Then we have

$$\sum_{t,s} \mathrm{Tr}A_t^2A_s^2 = \sum_{t,s} (\mathrm{Tr}A_tA_s)^2. \tag{5.1}$$

Proof. Using (1.2), we have

$$\sum_{t,s} \mathrm{Tr}A_t^2A_s^2 = \sum_{t,s,i,j,k,l} h^t_{kl}h^t_{li}h^s_{ij}h^s_{jk}$$

$$= \sum_{t,s,i,j,k,l} h^k_{tl}h^i_{lt}h^i_{sj}h^k_{js} = \sum_{k,i} (\mathrm{Tr}A_kA_i)^2.$$

Lemma 5.5. Let M be an n-dimensional anti-invariant submanifold of constant curvature k of a complex space form $\bar{M}^m(c)$. If the f-structure in the normal bundle is parallel, then we have

$$(\tfrac{1}{4}c-k)\sum_t [\mathrm{Tr}A_t^2 - (\mathrm{Tr}A_t)^2] = \sum_{t,s} [\mathrm{Tr}A_t^2A_s^2 - \mathrm{Tr}(A_tA_s)^2]. \tag{5.2}$$

Proof. From (4.6) of Chapter I and (2.7) we have

$$(\tfrac{1}{4}c-k)(\delta_{ik}\delta_{jl} - \delta_{il}\delta_{jk}) = \sum_t (h^t_{il}h^t_{jk} - h^t_{ik}h^t_{jl}). \tag{5.3}$$

Multiplying the both sides of (5.3) by $\sum_s h^t_{il}h^t_{jk}$, summing up with respect to i, j, k and l and using (5.1), we find (5.2).

Lemma 5.6. Let M be an n-dimensional anti-invariant submanifold of constant curvature k of a complex space form $\bar{M}^m(c)$. If the f-structure in the normal bundle is parallel, then we have

$$(5.4)\qquad (n-1)(\tfrac{1}{4}c-k)\sum_t \mathrm{Tr}A_t^2 = \sum_{t,s}[\mathrm{Tr}A_t^2A_s^2 - \mathrm{Tr}A_s\mathrm{Tr}(A_tA_sA_t)].$$

Proof. Putting i = k in (5.3) and summing up with respect to i, we have

$$(5.5)\qquad (n-1)(\tfrac{1}{4}c-k)\delta_{ji} = \sum_{t,i}(h_{il}^t h_{ij}^t - h_{ii}^t h_{jl}^t).$$

Multiplying the both sides of (5.5) by $\sum_s h_{jk}^s h_{kl}^s$ and summing up with respect to j, k and l we obtain (5.4).

§6. Anti-invariant submanifolds of constant curvature

Proposition 6.1. Let M be an n-dimensional anti-invariant submanifold with parallel mean curvature vector of a complex space form $\bar{M}^m(c)$. If M is of constant curvature k and moreover if the f-structure in the normal bundle is parallel, then we have

$$(6.1)\qquad \sum_{a,i,j,k}(h_{ijk}^a)^2 = -k\sum_t[(n+1)\mathrm{Tr}A_t^2 - 2(\mathrm{Tr}A_t)^2].$$

Proof. By the assumption and (1.5) the scalar curvature r of M is given by

$$r = \tfrac{1}{4}n(n-1)c + \sum_a[(\mathrm{Tr}A_a)^2 - \mathrm{Tr}A_a^2].$$

Since both r and $\sum_a(\mathrm{Tr}A_a)^2$ are constant, we see that the square of the length of the second fundamental form of M is also constant, that is, $S = \sum_a \mathrm{Tr}A_a^2$ = constant. Thus we see that

$$\sum_{a,i,j} h^a_{ij}\Delta h^a_{ij} = \frac{1}{2}\sum_a \mathrm{Tr}A^2_a - \sum_{a,i,j,k}(h^a_{ijk})^2 = - \sum_{a,i,j,k}(h^a_{ijk})^2.$$

Therefore (3.2) becomes

$$\sum_{a,i,j,k}(h^a_{ijk})^2 = - \sum_t[\tfrac{1}{4}(n+1)c\mathrm{Tr}A^2_t - \tfrac{1}{2}(\mathrm{Tr}A_t)^2] \tag{6.2}$$

$$- \sum_{t,s}\{\mathrm{Tr}(A_tA_s - A_sA_t)^2 - [\mathrm{Tr}(A_tA_s)]^2 + \mathrm{Tr}A_s\mathrm{Tr}(A_tA_sA_t)\}.$$

Substituting (5.2) and (5.4) into (6.2) and using (5.1), we have (6.1).

Proposition 6.2. Let M be an n-dimensional ($n > 1$) anti-invariant submanifold of a complex space form $\bar{M}^m(c)$ and M be with parallel mean curvature vector and of constant curvature k. If $\frac{1}{4}c \geq k$ and the f-structure in the normal bundle is parallel, then $k \leq 0$ or M is totally geodesic and $\frac{1}{4}c = k$.

Proof. From (5.5) we have

$$(\tfrac{1}{4}c-k)n(n-1) = \sum_t[\mathrm{Tr}A^2_t - (\mathrm{Tr}A_t)^2]. \tag{6.3}$$

Since $\frac{1}{4}c \geq k$, we have

$$\sum_t \mathrm{Tr}A^2_t \geq \sum_t(\mathrm{Tr}A_t)^2. \tag{6.4}$$

If $k > 0$, (6.1) implies that

$$0 = \sum_t(n-1)\mathrm{Tr}A^2_t + 2\sum_t[\mathrm{Tr}A^2_t - (\mathrm{Tr}A_t)^2],$$

from which we have $\sum_t \mathrm{Tr}A^2_t = 0$ and hence M is totally geodesic. In this case $\frac{1}{4}c = k$. Except for this possibility we have $k \leq 0$.

Proposition 6.3. Let M be an n-dimensional ($n > 1$) anti-invariant submanifold of a complex space form $\bar{M}^m(c)$ and M be with parallel second fundamental form and of constant curvature k. If $\frac{1}{4}c \geq k$ and the f-structure in the normal bundle is parallel, then either M is totally geodesic ($\frac{1}{4}c = k$) or M is flat ($k = 0$).

In Proposition 6.1, if M is minimal, we have

$$\sum_{a,i,j,k} (h^a_{ijk})^2 = - k(n+1)S,$$

and we have always $\frac{1}{4}c \geq k$ by (6.3). Consequently, we have the following

Theorem 6.1. Let M be an n-dimensional anti-invariant minimal submanifold of a complex space form $\bar{M}^m(c)$ and M be of constant curvature k. If the f-structure in the normal bundle is parallel, then either $k \geq 0$ or M is totally geodesic.

Theorem 6.2. Let M be an n-dimensional anti-invariant minimal submanifold of a complex space form $\bar{M}^m(c)$ and M be of constant curvature k and with parallel second fundamental form. If the f-structure in the normal bundle is parallel, then either M is totally geodesic or flat.

When n = m, Chen-Ogiue [6] proved the following

Corollary 6.1. Let M be an n-dimensional anti-invariant minimal submanifold of a complex space form $\bar{M}^n(c)$. If M is of constant curvature k, then either $k \leq 0$ or M is totally geodesic.

Corollary 6.2. Let M be an n-dimensional anti-invariant minimal submanifold of constant curvature k of a complex space form $\bar{M}^n(c)$. If the second fundamental forms of M are parallel, then either M is flat or M is totally geodesic.

Proposition 6.4. Let M be an n-dimensional anti-invariant submanifold with parallel mean curvature vector of a complex space form $\bar{M}^m(c)$. If the second fundamental forms of M are commutative and the f-structure in the normal bundle is parallel, then we have

$$(6.5) \qquad \sum_{a,i,j,k} (h^a_{ijk})^2 = -\frac{1}{4}c(n-1)\sum_t \mathrm{Tr}A_t^2 = -\frac{1}{4}c(n-1)S.$$

Proof. Using Lemmas 5.2 and 5.3, we can transform (6.1) into (6.5).

Proposition 6.5. Let M be an n-dimensional $(n > 1)$ anti-invariant submanifold of a complex space form $\bar{M}^m(c)$ and M be with parallel mean curvature vector and with commutative second fundamental forms. If the f-structure in the normal bundle is parallel, then either M is totally geodesic or $c \leq 0$.

Proposition 6.6. Let M be an n-dimensional $(n > 1)$ anti-invariant submanifold with parallel and commutative second fundamental forms of a complex space form $\bar{M}^m(c)$. If the f-structure in the normal bundle is parallel, then either M is totally geodesic or flat.

Proof. By the assumption and Lemma 5.3, M is of constant curvature $\frac{1}{4}c$. On the other hand, (6.5) implies that M is totally geodesic or $c = 0$ in which case M is flat.

Proposition 6.7. Let M be an n-dimensional anti-invariant submanifold with parallel mean curvature vector of a real 2m-dimensional flat Kaehlerian manifold $\bar{M}$. If the f-structure in the normal bundle is parallel and if the second fundamental forms of M are commutative, then the second fundamental form of M is parallel.

§7. Flat anti-invariant submanifolds

First of all, we notice that a simply connected complete Kaehlerian manifold of constant holomorphic sectional curvature c and of real dimension 2n can be identified with the complex projective space CP^n, the open unit ball D_n in C^n or C^n according as $c > 0$, $c < 0$ or $c = 0$. (See Chapter I, §6.)

Example 7.1. We now give an example of an anti-invariant submanifold in C^n. Let J be the complex structure of C^n given by

$$J = \begin{pmatrix} 0 & -1 & & & & & \\ 1 & 0 & & & & & \\ & & 0 & -1 & & & \\ & & 1 & 0 & & & \\ & & & & \ddots & & \\ & & & & & 0 & -1 \\ & & & & & 1 & 0 \end{pmatrix}.$$

Let $S^1(r_i) = \{z_i \in C : |z_i|^2 = r_i^2\}$, $i = 1,\ldots,n$, be circles of radius r_i. We consider

$$M^n = S^1(r_1) \times S^1(r_2) \times \ldots\ldots \times S^1(r_n)$$

in C^n, which is obviously flat. The position vector X of M^n in C^n has components given by

$$X = \begin{pmatrix} r_1 \cos u^1 \\ r_1 \sin u^1 \\ \cdot \\ \cdot \\ \cdot \\ \cdot \\ r_n \cos u^n \\ r_n \sin u^n \end{pmatrix}.$$

Putting $X_i = \partial_i X = \partial X/\partial u^i$, we have

$$X_1 = r_1 \begin{pmatrix} -\sin u^1 \\ \cos u^1 \\ 0 \\ \cdot \\ \cdot \\ \cdot \\ \cdot \\ 0 \end{pmatrix}, \ldots\ldots\ldots\ldots, X_n = r_n \begin{pmatrix} 0 \\ \cdot \\ \cdot \\ \cdot \\ \cdot \\ 0 \\ -\sin u^n \\ \cos u^n \end{pmatrix}.$$

On the other hand, we can take

$$N_1 = - \begin{pmatrix} \cos u^1 \\ \sin u^1 \\ 0 \\ \cdot \\ \cdot \\ \cdot \\ \cdot \\ 0 \end{pmatrix}, \ldots\ldots\ldots\ldots, N_n = - \begin{pmatrix} 0 \\ \cdot \\ \cdot \\ \cdot \\ \cdot \\ 0 \\ \cos u^n \\ \sin u^n \end{pmatrix}$$

as orthonormal vectors normal to M^n. Then we obtain

$$JX_1 = r_1 N_1, \quad JX_2 = r_2 N_2, \ldots\ldots\ldots\ldots, JX_n = r_n N_n.$$

Consequently M^n is a flat anti-invariant submanifold in C^n and it has parallel mean curvature vector and flat normal connection (see also [76], p.111). In view of Lemma 5.3 and Proposition 6.7, we see that M^n has parallel and commutative second fundamental forms. On the other hand, C^n is a totally geodesic Kaehlerian submanifold of C^m $(m > n)$. Thus M^n is an anti-invariant submanifold of C^m with parallel f-structure in the normal bundle.

Example 7.2. As an example similar to Example 7.1, we consider

$$M^n = S^1(r_1) \times \ldots\ldots \times S^1(r_p) \times R^{n-p}, \quad 1 \leq p < n.$$

Obviously R^{n-p} is a totally geodesic anti-invariant submanifold of C^{n-p}. Thus M^n is a flat complete anti-invariant submanifold of C^n. Moreover M^n is an anti-invariant submanifold with parallel and commutative second fundamental forms and with parallel f-structure of C^m $(m > n)$.

Theorem 7.1 (Yano-Kon [78]). Let M be an n-dimensional $(n > 1)$ complete anti-invariant submanifold of C^m and M be with parallel mean curvature vector and with commutative second fundamental forms. If the f-structure in the normal bundle is parallel, then either M is an n-dimensional plane R^n in a C^n in C^m, or a pythagorean product of the form

$$S^1(r_1) \times S^1(r_2) \times \ldots\ldots \times S^1(r_n) \quad \text{in a } C^n \text{ in } C^m,$$

or a pythagorean product of the form

$$S^1(r_1) \times \ldots\ldots \times S^1(r_p) \times R^{n-p} \quad \text{in a } C^n \text{ in } C^m,$$

where R^{n-p} is an (n-p)-dimensional plane and $1 \leq p < n$.

Proof. By the assumption and Lemma 5.3, M is flat. Thus Proposition 6.7 shows that the second fundamental forms of M are parallel. Moreover, using Lemma 5.1, we see that the normal connection of M is flat. From the fundamental theorem of submanifolds, Theorem 2.1 of Chapter II, we see that M is an anti-invariant submanifold of some C^n in C^m (see also Yano-Ishihara [76], Lemma 2.9). For a frame field, chosen as in Lemma 5.2, we define the following distributions:

$$T_t(x) = \{X \in T_x(M) : A_t X = \lambda_t X\} \quad \text{for } \lambda_t \neq 0,$$

$$T_0(x) = \{X \in T_x(M) : A_t X = 0,\ 1 \leq t \leq n\}.$$

Then each T_t is of dimension 1 and we have the decomposition:

$$T_x(M) = T_1(x) + \ldots\ldots + T_p(x) + T_0(x) \text{ (orthogonal direct sum).}$$

Moreover, since the normal connection of M in C^n is flat and the second fundamental forms of M are parallel, we see that each T_t is involutive and totally geodesic in M. Thus we conclude that

$$M = M_1 \times \ldots\ldots \times M_p \times M_0 \quad \text{(Riemannian product)},$$

where M_t and M_0 are the maximal integral submanifolds of T_t and T_0 respectively. Since M is complete, we have our assertion (see Theorem 3 in Yano-Ishihara [76]).

Theorem 7.2 (Yano-Kon [78]). Let M be an n-dimensional (n > 1) complete anti-invariant submanifold of a simply connected complete complex space form $\bar{M}^m(c)$ and M be with parallel and commutative second fundamental forms. If M is not totally geodesic and if the f-structure in the normal bundle is parallel, then M is a pythagorean product of the form

$$S^1(r_1) \times \ldots\ldots \times S^1(r_n) \qquad \text{in a } C^n \text{ in } C^m,$$

or a pythagorean product of the form

$$S^1(r_1) \times \ldots\ldots \times S^1(r_p) \times R^{n-p} \qquad \text{in a } C^n \text{ in } C^m,$$

where $1 \leq p < n$.

Proof. By the assumption and Proposition 6.6, we have c = 0. In this case we may consider that the ambient manifold $\bar{M}^m(c)$ is C^m. Thus Theorem 7.2 follows from Theorem 7.1.

Theorem 7.3 (Yano-Kon [78]). Under the same assumption as in Theorem 7.1, if M is compact, then M is a pythagorean product of the form

$$S^1(r_1) \times S^1(r_2) \times \ldots\ldots \times S^1(r_n) \qquad \text{in a } C^n \text{ in } C^m.$$

Theorem 7.4 (Yano-Kon [78]). Under the same assumption as in Theorem 7.2, if M is compact, then M is a pythagorean product of the form

$$S^1(r_1) \times S^1(r_2) \times \ldots\ldots \times S^1(r_n) \qquad \text{in a } C^n \text{ in } C^m.$$

When n = m, we have the following

Theorem 7.5. Let M be an n-dimensional (n > 1) complete anti-invariant submanifold of C^n and M be with parallel mean curvature vector and with commutative second fundamental forms. Then either M is an n-dimensional plane R^n, or a pythagorean product of the form

$$S^1(r_1) \times S^1(r_2) \times \ldots\ldots \times S^1(r_n) \qquad \text{in } C^n,$$

or a pythagorean product of the form

$$S^1(r_1) \times \ldots\ldots \times S^1(r_p) \times R^{n-p} \qquad \text{in } C^n, \quad 1 \leq p < n.$$

Theorem 7.6. Let M be an n-dimensional (n > 1) complete anti-invariant submanifold of a simply connected complete complex space form $\bar{M}^n(c)$ with parallel and commutative second fundamental forms. If M is not totally geodesic, then M is a pythagorean product of the form

$$S^1(r_1) \times S^1(r_2) \times \ldots\ldots \times S^1(r_n) \qquad \text{in } C^n,$$

or a pythagorean product of the form

$$S^1(r_1) \times \ldots\ldots \times S^1(r_p) \times R^{n-p} \qquad \text{in } C^n, \quad 1 \leq p < n.$$

Theorem 7.7 (Yano-Kon [77]). Under the same assumption as in Theorem 7.5, if M is compact, then M is a pythagorean product of the form

$$S^1(r_1) \times S^1(r_2) \times \ldots\ldots \times S^1(r_n) \qquad \text{in } C^n.$$

Theorem 7.8 (Yano-Kon [77]). Under the same assumption as in Theorem 7.6, if M is compact, then M is a pythagorean product of the form

$$S^1(r_1)\times S^1(r_2)\times \ldots\ldots \times S^1(r_n) \qquad \text{in } C^n.$$

§8. Scalar curvature

Let M be an n-dimensional anti-invariant minimal submanifold of a complex space form $\bar{M}^m(c)$ with parallel f-structure in the normal bundle. Then from (1.2), (2.7), (2.10) and (2.11) we have

$$(8.1) \qquad \sum_{t,s}(R_{ts})^2 = \frac{1}{n}r^2 - \frac{1}{n}S^2 + \sum_{t,s}\mathrm{Tr}A_t^2A_s^2,$$

$$(8.2) \qquad \sum_{i,j,k,l}(R^i_{jkl})^2 = \frac{2}{n(n-1)}r^2 - \frac{2}{n(n-1)}S^2 - \sum_{t,s}\mathrm{Tr}(A_tA_s - A_sA_t)^2,$$

where r denotes the scalar curvature and S the square of the length of the second fundamental form of M respectively. We put

$$K_N = \sum_{a,b}\mathrm{Tr}(A_aA_b - A_bA_a)^2.$$

Since f-structure in the normal bundle is parallel, we have

$$K_N = \sum_{t,s}\mathrm{Tr}(A_tA_s - A_sA_t)^2.$$

K_N is called the scalar normal curvature of M. The scalar normal curvature K_N of anti-invariant surfaces of CP^2 are studied by Houh [15]. (See section 9, Theorem 9.1.)

We have in general

$$\sum_{t,s}(R_{ts})^2 \geq \frac{1}{n}r^2, \qquad \sum_{i,j,k,l}(R^i_{jkl})^2 \geq \frac{2}{n(n-1)}r^2.$$

In the first inequality the equality holds if and only if M is an Einstein manifold when $n \geq 2$ and in the second inequality the equality holds if and only if M is of constant curvature when $n \geq 3$. Therefore (8.2) implies the following

Proposition 8.1. Let M be an n-dimensional ($n \geq 3$) anti-invariant minimal submanifold with parallel f-structure in the normal bundle of a complex space form $\bar{M}^m(c)$. Then $K_N \geq 2S^2/n(n-1)$ and equality holds if and only if M is of constant curvature.

Now we assume that the second fundamental forms of M are parallel, then (3.4) and (5.1) imply

$$\frac{1}{4}(n+1)cS + \sum_{t,s} [\mathrm{Tr}(A_tA_s - A_sA_t)^2 - \mathrm{Tr}A_t^2A_s^2] = 0. \tag{8.3}$$

If $c \leq 0$, then (8.3) shows that M is totally geodesic. Thus we have

Proposition 8.2. Let M be an n-dimensional anti-invariant minimal submanifold with parallel second fundamental forms and with parallel f-structure in the normal bundle of a complex space form $\bar{M}^m(c)$. If $c \leq 0$, then M is totally geodesic.

Next we prove the following

Theorem 8.1. Let M be an n-dimensional anti-invariant minimal submanifold with parallel f-structure in the normal bundle of a complex space form $\bar{M}^m(c)$. Then either M is totally geodesic or M has the non-negative scalar curvature $r \geq 0$. Moreover if $r = 0$, then M is flat.

Proof. From (8.1), (8.2) and (8.3) we have

$$\frac{(n+1)}{n(n-1)} Sr = \sum_{i,j,k,l} (R^i_{jkl})^2 - \frac{2}{n(n-1)} r^2 + \sum_{i,j} (R_{ij})^2 - \frac{1}{n}r^2 \geq 0. \tag{8.4}$$

Consequently, either $S = 0$, i.e., M is totally geodesic or $r \geq 0$. Moreover, if $r = 0$, then M is obviously flat by (8.4).

Corollary 8.1 (Kon [25]). Let M be an n-dimensional anti-invariant minimal submanifold with parallel second fundamental forms of a complex space form $\bar{M}^n(c)$. Then either M is totally geodesic or M has non-negative scalar curvature $r \geq 0$. Moreover, if $r = 0$, then M is flat.

§9. Anti-invariant surfaces

Let M be an anti-invariant minimal surface of a real 4-dimensional complex space form $\bar{M}^2(c)$. By a suitable choice of a frame, the second fundamental forms of M are expressed by the following matrices:

$$(9.1) \qquad A_1 = \begin{pmatrix} a & 0 \\ 0 & -a \end{pmatrix}, \quad A_2 = \begin{pmatrix} 0 & -a \\ -a & 0 \end{pmatrix}.$$

Then the scalar normal curvature K_N of M is given by

$$(9.2) \qquad K_N = -16a^4.$$

Thus if the scalar normal curvature K_N of M is constant, then a is a constant. Thus (1.7) shows that M is of constant curvature. From Corollary 6.1 we have the following theorem proved by Houh [15].

Theorem 9.1. If M is an anti-invariant minimal surface with constant scalar normal curvature of $\bar{M}^2(c)$, then M is either totally geodesic or M is a surface with non-positive curvature.

To close this section, we prove the following

Theorem 9.2 (Yau [82]). Let M be an anti-invariant minimal surface of a complex space form $\bar{M}^2(c)$.

(1) If M has genus zero, then M is totally geodesic.

(2) If M is a complete surface with non-negative curvature, then M is totally geodesic or flat.

(3) If M is a complete surface with non-positive Gauss curvature K and if $\frac{1}{4}c - K \geq \alpha > 0$ for some constant α, then M is flat.

Proof. From (3.4) and (9.1) we have

$$\sum_{t,i,j} h^t_{ij}\Delta h^t_{ij} = 3ca^2 - 24a^4. \tag{9.3}$$

On the other hand, by (2.7) the Gauss curvature K of M is given by

$$K = \frac{1}{4}c - 2a^2. \tag{9.4}$$

From (9.3) and (9.4) we have

$$\sum_{t,i,j} h^t_{ij}\Delta h^t_{ij} = 12a^2K. \tag{9.5}$$

Since $\sum_{t,i,j} h^t_{ij}\Delta h^t_{ij} = \frac{1}{2}\Delta S - \sum_{t,i,j,k} (h^t_{ijk})^2$, we have

$$\frac{1}{2}\Delta S = \sum_{t,i,j,k} (h^t_{ijk})^2 + 12a^2K. \tag{9.6}$$

From this we see that

$$\Delta \log S = 6K. \tag{9.7}$$

Using the isothermal coordinates and (9.7), we see that S = 0 or S vanishes only at isolated points. If M has genus zero, the Gauss-Bonnet theorem and (9.7) show that M is totally geodesic.

If M is complete and has non-negative curvature, (9.6) shows that either M is totally geodesic or flat.

In case M is complete and has non-positive curvature, by (9.7) the curvature of the metric $S^{2/3}ds^2$ is zero. If S is bounded from below by zero, this metric is complete and hence M is parabolic. Since (9.7) shows that log S is a non-negative superharmonic function and hence K = 0. By (9.4) and the assumption S is indeed bounded from below by zero.

§10. Anti-invariant submanifolds of a Kaehlerian manifold with vanishing Bochner curvature tensor

In this section we study anti-invariant submanifolds of Kaehlerian manifolds with vanishing Bochner curvature tensor.

Let $\bar{M}$ be a real 2m-dimensional Kaehlerian manifold. We now consider the so called Bochner curvature tensor (see Tachibana [55] and Yano-Bochner [73]) defined by

$$(10.1)\qquad \begin{aligned} B_{ABCD} = K_{ABCD} &+ \delta_{AC}L_{BD} - \delta_{AD}L_{BC} + L_{AC}\delta_{BD} - L_{AD}\delta_{BC} \\ &+ J_{AC}M_{BD} - J_{AD}M_{BC} + M_{AC}J_{BD} - M_{AD}J_{BC} \\ &+ 2(M_{AB}J_{CD} + J_{AB}M_{CD}), \end{aligned}$$

where we have put

$$(10.2)\qquad L_{BD} = -\frac{1}{2(m+2)}K_{BD} + \frac{1}{8(m+1)(m+2)}K\delta_{BD},$$

$$(10.3)\qquad M_{BD} = -\sum_C L_{BC}J_{DC},$$

where K_{BD} and K denote the Ricci tensor and the scalar curvature of $\bar{M}$ respectively. We see that $L_{BD} = L_{DB}$ and $M_{BD} = -M_{DB}$.

By a straightforward computation we have (Tachibana [55])

$$(10.4)\qquad \sum_A \nabla_A B_{ABCD} = -2m[\nabla_C L_{BD} - \nabla_D L_{CB} + \frac{1}{8(m+1)(m+2)} \sum_A (J_{CA}J_{DB} - J_{DA}J_{CB} - 2J_{CD}J_{BA})(\nabla_A K)].$$

Let M be an n-dimensional anti-invariant submanifold of $\bar{M}$. Then by (10.1) we find

$$(10.5)\qquad B_{ijkl} = K_{ijkl} + \delta_{ik}L_{jl} - \delta_{il}L_{jk} + L_{ik}\delta_{jl} - L_{il}\delta_{jk}.$$

In the following, we rewrite the Gauss equation (3.7) of Chapter II as follows:

$$(10.6)\qquad K_{ijkl} = R_{ijkl} - D_{ijkl},$$

where we have put

$$(10.7)\qquad D_{ijkl} = \sum_a (h^a_{ik}h^a_{jl} - h^a_{il}h^a_{jk}).$$

Hereafter we put

$$D_{ik} = \sum_j D_{ijkj}, \qquad D = \sum_i D_{ii}, \qquad b_{ik} = \sum_j B_{ijkj},$$
$$b = \sum_i b_{ii}, \qquad G = \sum_i L_{ii}.$$

From (10.5) and (10.6) we have

$$(10.8)\qquad B_{ijkl} = R_{ijkl} - D_{ijkl} + \delta_{ik}L_{jl} - \delta_{il}L_{jk} + L_{ik}\delta_{jl} - L_{il}\delta_{jk}.$$

Consequently we obtain, by contraction, the following equations:

$$(10.9)\qquad b_{jl} = R_{jl} - D_{jl} + (n-2)L_{jl} + G\delta_{jl},$$

$$(10.10)\qquad b = r - D + 2(n-1)G,$$

where R_{ij} and r denote the Ricci tensor and the scalar curvature of M respectively. From (10.9) and (10.10) we find

$$L_{jl} = -\frac{1}{(n-2)}(R_{jl} - D_{jl} - b_{jl}) + \frac{1}{2(n-1)(n-2)}(r - D - b)\delta_{jl}. \tag{10.11}$$

Substituting (10.11) into (10.8) we have

$$\begin{aligned} B_{ijkl} = C_{ijkl} - D_{ijkl} + \frac{1}{(n-2)}(&\delta_{ik}b_{jl} + \delta_{jl}b_{ik} - \delta_{il}b_{jk} \\ &- \delta_{jk}b_{il} + \delta_{ik}D_{jl} + \delta_{jl}D_{ik} - \delta_{il}D_{jk} - \delta_{jk}D_{il}) \\ &- \frac{1}{(n-1)(n-2)}(\delta_{ik}\delta_{jl} - \delta_{il}\delta_{jk})(b + D), \end{aligned} \tag{10.12}$$

where C_{ijkl} denotes the Weyl conformal curvature tensor.

Lemma 10.1. Let M be an n-dimensional ($n \geq 4$) anti-invariant submanifold of a real 2m-dimensional Kaehlerian manifold $\bar{M}$ with vanishing Bochner curvature tensor. If

$$D_{ijkl} = \alpha(\delta_{ik}\delta_{jl} - \delta_{il}\delta_{jk}) \tag{10.13}$$

for some scalar function α on M, then M is conformally flat.

Proof. Since $\bar{M}$ has vanishing Bochner curvature tensor, we have

$$B_{ijkl} = 0, \qquad b_{jl} = 0 \qquad \text{and} \qquad b = 0.$$

Thus (10.12) becomes

$$\begin{aligned} C_{ijkl} = D_{ijkl} - \frac{1}{(n-2)}&(\delta_{ik}D_{jl} + \delta_{jl}D_{ik} - \delta_{il}D_{jk} - \delta_{jk}D_{il}) \\ &+ \frac{1}{(n-1)(n-2)}(\delta_{ik}\delta_{jl} - \delta_{il}\delta_{jk})D. \end{aligned} \tag{10.14}$$

On the other hand, by the assumption (10.13) we find, by contraction,

$$D_{jl} = \alpha(n-1)\delta_{jl}, \qquad D = \alpha n(n-1). \tag{10.15}$$

Substituting (10.15) into (10.14) we have $C_{ijkl} = 0$ and consequently M is conformally flat.

If M is totally umbilical. then we have

$$D_{ijkl} = |m|^2(\delta_{ik}\delta_{jl} - \delta_{il}\delta_{jk}) \tag{10.16}$$

where m is the mean curvature vector of M and $|m|^2 = \sum_a (\mathrm{Tr}A_a)^2/n^2$, which is globally defined on M. From this and Lemma 10.1 we have

Theorem 10.1 (Yano [69]). Let M be an n-dimensional ($n \geq 4$) totally umbilical, anti-invariant submanifold of a real 2m-dimensional Kaehlerian manifold $\bar{M}$ with vanishing Bochner curvature tensor. Then M is conformally flat.

When the second fundamental forms of M are commutative, we have the following

Theorem 10.2 (Yano [70]). Let M be an n-dimensional ($n \geq 4$) anti-invariant submanifold of a real 2m-dimensional Kaehlerian manifold $\bar{M}$ with vanishing Bochner curvature tensor. If the second fundamental forms of M are commutative and if the f-structure in the normal bundle is parallel, then M is conformally flat.

Proof. From the assumption we have

$$D_{ijkl} = \sum_t (h^t_{ik}h^t_{jl} - h^t_{il}h^t_{jk}) = \sum_t (h^i_{tk}h^j_{tl} - h^i_{tl}h^j_{tk}) = 0.$$

Thus Lemma 10.1 proves our assertion.

Corollary 10.1. Let M be an n-dimensional ($n \geq 4$) anti-invariant submanifold of a real 2n-dimensional Kaehlerian manifold $\bar{M}$ with vanishing Bochner curvature tensor. If the second fundamental forms of M are commutative, then M is conformally flat.

In the following we study an anti-invariant submanifold of dimension 3. We prove the following

Theorem 10.3 (Yano [70]). Let M be a 3-dimensional anti-invariant submanifold of a real 2m-dimensional Kaehlerian manifold $\bar{M}$ with vanishing Bochner curvature tensor. If M is totally umbilical, then M is conformally flat.

Proof. We put $H_a = (\mathrm{Tr}A_a)/n$. Then (10.11), by the assumption, becomes

$$L_{jl} = C_{jl} + \frac{1}{2}\sum_a H_a^2\delta_{jl}, \tag{10.17}$$

where

$$C_{jl} = -\frac{1}{(n-2)}R_{jl} + \frac{1}{2(n-1)(n-2)}r\delta_{jl}.$$

On the other hand, the Codazzi equation (2.5) of Chapter II implies

$$\sum_i K_{aiji} = \sum_i (h^a_{iij} - h^a_{iji}) = (n-1)\nabla_j H_a. \tag{10.18}$$

By (10.1) we also have

$$\sum_i K_{aiji} = -(n-1)L_{aj} + 3\sum_{C,i} J_{ai}L_{jC}J_{iC}. \tag{10.19}$$

From (10.18) and (10.19) we have

$$\nabla_j H_a = -L_{aj} + \frac{3}{(n-1)}\sum_{C,i} J_{ai}L_{jC}J_{iC}. \tag{10.20}$$

Since M is umbilical, the second fundamental forms of M are commutative and by Lemma 5.2 (see proof of Corollary 5.2) we see that $\sum_a J_{ai}H_a = H_{i*} = 0$. Thus we have

$$\sum_a (\nabla_j H_a)H_a = -\sum_a L_{aj}H_a. \tag{10.21}$$

From (10.17) we have

$$\bar{\nabla}_i L_{jl} + \sum_a h^a_{ij} L_{al} + \sum_a k^a_{il} L_{ja} = \nabla_i C_{jl} + \sum_a (\nabla_i H_a) H_a \delta_{jl}. \tag{10.22}$$

Since M is umbilical, by (10.22) we have

$$\begin{aligned} \bar{\nabla}_i L_{jl} - \bar{\nabla}_j L_{il} + \sum_a H_a L_{ja} \delta_{il} - \sum_a H_a L_{ia} \delta_{jl} \\ = \nabla_i C_{jl} - \nabla_j C_{il} + \sum_a (\nabla_i H_a) H_a \delta_{jl} - \sum_a (\nabla_j H_a) H_a \delta_{il}. \end{aligned} \tag{10.23}$$

From (10.21) and (10.23) we find that

$$\bar{\nabla}_i L_{jl} - \bar{\nabla}_j L_{il} = \nabla_i C_{jl} - \nabla_j C_{il}. \tag{10.24}$$

By the assumption and (10.4) we have

$$\nabla_i C_{jl} - \nabla_j C_{il} = 0,$$

which proves that M is conformally flat.

Corollary 10.2 (Blair [3]). Let M be an n-dimensional ($n \geq 3$) anti-invariant submanifold of a real 2m-dimensional Kaehlerian manifold $\bar{M}$ with vanishing Bochner curvature tensor. If M is totally geodesic, then M is conformally flat.

Chapter IV

Anti-invariant submanifolds of Sasakian manifolds tangent to the structure vector field

In this chapter, we shall study anti-invariant submanifolds of Sasakian manifolds tangent to the structure vector field. Therefore, in this chapter, we mean by an anti-invariant submanifold M of a Sasakian manifold $\bar{M}$ an anti-invariant submanifold M of $\bar{M}$ tangent to the structure vector field ξ of $\bar{M}$.

§1. Second fundamental forms

Let $\bar{M}$ be a Sasakian manifold of dimension 2m+1 with structure tensors $(\phi,\xi,\eta,\bar{g})$ and let M be an (n+1)-dimensional anti-invariant submanifold of $\bar{M}$. Then for any tangent vector field X to M we have

$$\phi X = \bar{\nabla}_X\xi = \nabla_X\xi + B(X,\xi),$$

where $\bar{\nabla}$ (resp. ∇) denotes the operator of covariant differentiation with respect to the Levi-Civita connection in $\bar{M}$ (resp. M) and B the second fundamental form of M. Since M is anti-invariant in $\bar{M}$ and $\nabla_X\xi = 0$ (Proposition 7.2 of Chapter II), we have

$$\phi X = B(X,\xi) \qquad \text{and} \qquad B(\xi,\xi) = 0.$$

Proposition 1.1. Let $\bar{M}$ be a Sasakian manifold of dimension 2m+1 and M be an anti-invariant submanifold of $\bar{M}$ of dimension n+1. If $n \geq 1$, then M is not totally umbilical.

Proof. If M is totally umbilical, then we have $B(X,Y) = g(X,Y)m$ for any vector fields X and Y, where m denotes the mean curvature vector of M. Since $B(\xi,\xi) = 0$, we have $g(\xi,\xi)m = 0$ and hence $m = 0$, which shows that M is minimal submanifold. Therefore M is totally geodesic, which is a contradiction to the fact that $\phi X = B(X,\xi) \neq 0$.

We choose a local field of orthonormal frames $e_0=\xi, e_1,\ldots,e_n; e_{n+1},\ldots, e_m; e_{1*}=\phi e_1,\ldots,e_{n*}=\phi e_n; e_{(n+1)*}=\phi e_{n+1},\ldots,e_{m*}=\phi e_m$ in $\bar{M}$ in such a way that, restricted to M, $e_0,e_1,\ldots,e_n$ are tangent to M. With respect to this frame field of $\bar{M}$, let $\omega^0=\eta,\omega^1,\ldots,\omega^n;\omega^{n+1},\ldots,\omega^m;\omega^{1*},\ldots,\omega^{n*};\omega^{(n+1)*},\ldots,\omega^{m*}$ be the dual forms. Unless otherwise stated, we use the conventions that the ranges of indices are respectively:

$$A, B, C, D = 0,1,\ldots,m,1^*,\ldots,m^*,$$
$$i, j, k, l, t, s = 0,1,\ldots,n,$$
$$x, y, z, v, w = 1,\ldots,n,$$
$$a, b, c, d = n+1,\ldots,m,1^*,\ldots,m^*,$$
$$\lambda, \mu, \nu = n+1,\ldots,m,(n+1)^*,\ldots,m^*,$$
$$\alpha, \beta, \gamma = n+1,\ldots,m.$$

Then we have the following equations:

$$\begin{aligned}
&\omega^x_y = \omega^{x^*}_{y^*}, \quad \omega^{x^*}_y = \omega^{y^*}_x, \quad \omega^x = \omega^{x^*}_0, \quad \omega^{x^*} = -\omega^x_0,\\
(1.1)\qquad &\omega^\alpha_\beta = \omega^{\alpha^*}_{\beta^*}, \quad \omega^{\alpha^*}_\beta = \omega^{\beta^*}_\alpha, \quad \omega^\alpha = \omega^{\alpha^*}_0, \quad \omega^{\alpha^*} = -\omega^\alpha_0,\\
&\omega^x_\alpha = \omega^{x^*}_{\alpha^*}, \quad \omega^{x^*}_\alpha = \omega^{\alpha^*}_x.
\end{aligned}$$

Restricting these forms to M, we have $\omega^a = 0$ and hence $\omega^x_0 = \omega^{\alpha^*}_0 = \omega^\alpha_0 = 0$. Thus (3.4) of Chapter II and (1.1) imply that

(1.2) $\quad h^x_{yz} = h^y_{xz} = h^z_{xy}, \qquad h^\lambda_{0i} = 0, \qquad h^x_{00} = 0, \qquad h^x_{0i} = \delta_{xi},$

where we use h^x_{ij} in place of $h^{x^*}_{ij}$ to simplify the notation. For each a, the second fundamental form A_a are represented by the symmetric (n+1,n+1)-matrices $A_a = (h^a_{ij})$. From (1.2) we have

$$A_x = x\left(\begin{array}{c|c} 0 & 0\ldots\ldots 0\overset{x}{1}0\ldots\ldots 0 \\ \hline 0 & \\ \vdots & \\ 0 & \\ 1 & h^x_{yz} \\ 0 & \\ \vdots & \\ 0 & \end{array}\right), \qquad \text{for all } x, \tag{1.3}$$

$$A_\lambda = \left(\begin{array}{c|c} 0 & 0 \\ \hline & \\ 0 & h^\lambda_{yz} \\ & \end{array}\right), \qquad \text{for all } \lambda.$$

Hereafter, we put $H_a = (h^a_{yz})$, which are symmetric (n,n)-matrices. We denote by S the square of the length of the second fundamental form A of M, i.e., $S = \sum_{a,i,j} (h^a_{ij})^2 = \sum_a \mathrm{Tr} A_a^2$. On the other hand, we put

$$T = \sum_{a,x,y} (h^a_{xy})^2 = \sum_a \mathrm{Tr} H_a^2.$$

Then, by (1.3) we have

(1.4) $$S = T + 2n.$$

Moreover, we see that

$$\mathrm{Tr}A_a = \sum_i h^a_{ii} = \sum_x h^a_{xx} = \mathrm{Tr}H_a.$$

Thus M is minimal if and only if $\mathrm{Tr}H_a = 0$ for all a.

In the sequel, we study the normal bundle of an anti-invariant submanifold M of $\bar{M}$. Since $\phi T_x(M) \subset T_x(M)^\perp$ at each point x of M, we have the decomposition

$$T_x(M)^\perp = \phi T_x(M) \oplus N_x(M),$$

where $N_x(M)$ is an orthogonal complement of $\phi T_x(M)$ in the normal space $T_x(M)^\perp$ at x. If a vector $N \in N_x(M)$, then we have $\phi N \in N_x(M)$. If N is a vector field in the normal bundle $T(M)^\perp$, we put

(1.5) $$\phi N = PN + fN,$$

where PN is the tangential part of ϕN and fN the normal part of ϕN. Then P is a tangent bundle valued 1-form on the normal bundle and f is an endomorphism of the normal bundle. Putting $N = \phi X$ in (1.5) and applying ϕ to (1.5), we find

(1.6)
$$PfN = 0, \qquad f^2N = -N - \phi PN,$$
$$P\phi X = -X + \eta(X)\xi, \qquad f\phi X = 0$$

for any tangent vector field X to M and a normal vector field N to M. Equations (1.6) imply

$$f^3 + f = 0.$$

Since f is of constant rank, if f does not vanish, it defines an f-structure in the normal bundle. From (1.5), using the Gauss and Weingarten formulas, we have

(1.7) $$(D_Xf)N = -B(X,PN) - \phi A_N X.$$

If $D_X f = 0$ for all X, then the f-structure in the normal bundle is said to be parallel.

Lemma 1.1. Let M be an (n+1)-dimensional anti-invariant submanifold of a (2m+1)-dimensional Sasakian manifold $\bar{M}$. If the f-structure in the normal bundle is parallel, then we have

$$A_N = 0 \qquad \text{for} \qquad N \in N_x(M). \tag{1.8}$$

Proof. If $N \in N_x(M)$, then $PN = 0$. Thus by the assumption and (1.7) we have $\phi A_N X = 0$, from which

$$\phi^2 A_N X = -A_N X + \eta(A_N X)\xi = 0.$$

On the other hand, we can verify that

$$\eta(A_N X) = -\bar{g}(\bar{\nabla}_X N, \xi) = g(N, \phi X) = 0,$$

and hence $A_N X = 0$.

Remark. We can choose a frame $e_{n+1}, \ldots, e_m, e_{(n+1)^*}, \ldots, e_{m^*}$ for $N_x(M)$. Thus (1.8) can be rewritten as

$$A_\lambda = 0, \qquad \text{i.e.,} \qquad h^\lambda_{ij} = 0. \tag{1.9}$$

Moreover, by (1.7), if $H_a = 0$ for all a, then the f-structure in the normal bundle is parallel.

§2. Locally product anti-invariant submanifolds

Let M be an (n+1)-dimensional anti-invariant submanifold of a (2m+1)-dimensional Sasakian manifold $\bar{M}$. We denote by K and R the Riemannian curvature tensors of $\bar{M}$ and M respectively. Since M is an anti-invariant submanifold tangent to the structure vector field ξ of $\bar{M}$, (7.1) of Chapter I implies

$$K^{x^*}_{y^*ij} = K^x_{yij} - (\delta_{ix}\delta_{jy} - \delta_{iy}\delta_{jx}). \tag{2.1}$$

On the other hand, from Proposition 7.2 of Chapter II we have

$$R^0_{ijk} = R^i_{0jk} = 0. \tag{2.2}$$

If $\bar{M}$ is of constant ϕ-sectional curvature c, then we have

$$\begin{aligned} K^A_{BCD} = {} & \tfrac{1}{4}(c+3)(\delta_{AC}\delta_{BD} - \delta_{AD}\delta_{BC}) + \tfrac{1}{4}(\eta_B\eta_C\delta_{AD} - \eta_B\eta_D\delta_{AC} \\ & + \eta_A\eta_D\delta_{BC} - \eta_A\eta_C\delta_{BD} + \phi_{AC}\phi_{BD} - \phi_{AD}\phi_{BC} + 2\phi_{AB}\phi_{CD}), \end{aligned} \tag{2.3}$$

from which

$$\begin{aligned} K^i_{jkl} = {} & \tfrac{1}{4}(c+3)(\delta_{ik}\delta_{jl} - \delta_{il}\delta_{jk}) + \tfrac{1}{4}(c-1)(\eta_j\eta_k\delta_{il} - \eta_j\eta_l\delta_{ik} \\ & + \eta_i\eta_l\delta_{jk} - \eta_i\eta_k\delta_{jl}). \end{aligned} \tag{2.4}$$

Thus the Gauss equation (3.7) of Chapter II implies

$$\begin{aligned} R^i_{jkl} = {} & \tfrac{1}{4}(c+3)(\delta_{ik}\delta_{jl} - \delta_{il}\delta_{jk}) + \tfrac{1}{4}(c-1)(\eta_j\eta_k\delta_{il} - \eta_j\eta_l\delta_{ik} \\ & + \eta_i\eta_l\delta_{jk} - \eta_i\eta_k\delta_{jl}) + \sum_a(h^a_{ik}h^a_{jl} - h^a_{il}h^a_{jk}), \end{aligned} \tag{2.5}$$

from which we find that the Ricci tensor R_{ij} of M is given by

$$\begin{aligned} R_{ij} = {} & \tfrac{1}{4}[n(c+3)-(c-1)]\delta_{ij} - \tfrac{1}{4}(n-1)(c-1)\eta_i\eta_j \\ & + \sum_{a,k}(h^a_{kk}h^a_{ij} - h^a_{ik}h^a_{jk}). \end{aligned} \tag{2.6}$$

On the other hand, by (1.2) we have

$$\sum_{a,k}(h^a_{kk}h^a_{ij} - h^a_{ik}h^a_{jk}) = \sum_{a,x}(h^a_{xx}h^a_{ij} - h^a_{ix}h^a_{jx}) - \delta_{ij} + \eta_i\eta_j.$$

Thus (2.6) can be rewritten as

$$(2.7) \qquad R_{ij} = \frac{1}{4}(n-1)(c+3)\delta_{ij} - \frac{1}{4}[(n-1)c-(n+3)]\eta_i\eta_j + \sum_{a,x}(h^a_{xx}h^a_{ij} - h^a_{ix}h^a_{jx}).$$

Since $\sum_{a,i,x}(h^a_{ix})^2 = \sum_{a,x,y}(h^a_{xy})^2 + n$, we see from (1.2) that the scalar curvature of M is given by

$$(2.8) \qquad r = \frac{1}{4}n(n-1)(c+3) + \sum_{a,x,y}(h^a_{xx}h^a_{yy} - h^a_{xy}h^a_{xy}).$$

If M is minimal, then from (2.7) and (2.8) we have respectively

$$(2.9) \qquad R_{ij} = \frac{1}{4}(n-1)(c+3)\delta_{ij} - \frac{1}{4}[(n-1)c-(n+3)]\eta_i\eta_j - \sum_{a,x}h^a_{ix}h^a_{jx},$$

and

$$(2.10) \qquad r = \frac{1}{4}n(n-1)(c+3) - \sum_{a,x,y}(h^a_{xy})^2.$$

If the f-structure in the normal bundle is parallel, then (2.7) and (2.8) can be respectively rewritten as

$$(2.11) \qquad R_{ij} = \frac{1}{4}(n-1)(c+3)\delta_{ij} - \frac{1}{4}[(n-1)c-(n+3)]\eta_i\eta_j + \sum_{t,x}(h^t_{xx}h^t_{ij} - h^t_{ix}h^t_{jx}),$$

and

$$(2.12) \qquad r = \frac{1}{4}(n-1)(c+3) + \sum_{t,x,y}(h^t_{xx}h^t_{yy} - h^t_{xy}h^t_{xy}).$$

If M is moreover minimal, then we have

$$(2.13) \qquad R_{ij} = \frac{1}{4}(n-1)(c+3)\delta_{ij} - \frac{1}{4}[(n-1)c-(n+3)]\eta_i\eta_j - \sum_{t,x} h^t_{ix}h^t_{jx},$$

$$(2.14) \qquad r = \frac{1}{4}n(n-1)(c+3) - \sum_{t,x,y}(h^t_{xy})^2.$$

Proposition 2.1. Let M be an (n+1)-dimensional anti-invariant minimal submanifold of a Sasakian space form $\bar{M}^{2m+1}(c)$. Then for the Ricci tensor S and the scalar curvature r of M

(a) $S - \frac{1}{4}(n-1)(c+3)g + \frac{1}{4}[(n-1)c-(n+3)]\eta\otimes\eta$ is negative semi-definite,

(b) $r \leq \frac{1}{4}n(n-1)(c+3)$.

Proposition 2.2. Let $\bar{M}$ be a (2m+1)-dimensional Sasakian manifold and M be an (n+1)-dimensional anti-invariant submanifold of $\bar{M}$. If the f-structure in the normal bundle is parallel, then M is flat if and only if the normal connection of M is flat.

Proof. From (3.9) of Chapter II, we see that (1.2) and (2.1) imply

$$(2.15) \qquad R^{x^*}{}_{y^*ij} = K^{x^*}{}_{y^*ij} + \sum_k(h^x_{ki}h^y_{kj} - h^x_{kj}h^y_{ki})$$

$$= K^x_{yij} + \sum_z(h^z_{ix}h^z_{jy} - h^z_{jx}h^z_{ij}).$$

Since $h^\lambda_{ij} = 0$, by (3.7) of Chapter II, we have

$$(2.16) \qquad R^x_{yij} = K^x_{yij} + \sum_z(h^z_{ix}h^z_{jy} - h^z_{iy}h^z_{jx}).$$

From (2.15) and (2.16) we have $R^{x^*}{}_{y^*ij} = R^x_{yij}$. This combined with (2.2) proves our assertion.

Proposition 2.3. Let M be an (n+1)-dimensional ($n > 1$) anti-invariant submanifold with commutative second fundamental forms of a Sasakian space form $\bar{M}^{2m+1}(c)$. If the f-structure in the normal bundle is parallel, then M is flat if and only if $\bar{M}^{2m+1}(c)$ is of constant curvature 1, that is, $c = 1$.

Proof. From (1.2) and (1.8), we have, by a straightforward computation,

$$\sum_a (h^a_{ik}h^a_{jl} - h^a_{il}h^a_{jk}) = \sum_x (h^x_{ik}h^x_{jl} - h^x_{il}h^x_{jk}) = -(\delta_{ik}\delta_{jl} - \delta_{il}\delta_{jk}). \tag{2.16}$$

From (2.5) and (2.16) we find

$$R^i_{jkl} = \frac{1}{4}(c-1)(\delta_{ik}\delta_{jl} - \delta_{il}\delta_{jk} + \eta_j\eta_k\delta_{il} - \eta_j\eta_l\delta_{ik} + \eta_i\eta_l\delta_{jk} - \eta_i\eta_k\delta_{jl}), \tag{2.17}$$

which proves our assertion.

By Proposition 7.2 of Chapter II, the structure vector field ξ is parallel with respect to the induced connection ∇ on M. Therefore, by (2.17) and a theorem of Yano ([65], p.274), we have

Theorem 2.1. Let M be an (n+1)-dimensional anti-invariant submanifold with commutative second fundamental forms of a Sasakian space form $\bar{M}^{2m+1}(c)$. If the f-structure in the normal bundle is parallel, then M is locally a Riemannian direct product $M^n \times R^1$, where M^n is a hypersurface of M with constant curvature $\frac{1}{4}(c-1)$ and is totally geodesic in M.

Corollary 2.1. Under the same assumptions as those in Theorem 2.1, if $c = 1$, then M is locally a Riemannian direct product $M^n \times R^1$, where M^n is a flat hypersurface of M.

§3. Laplacian

Let $\bar{M}$ be a (2m+1)-dimensional Sasakian manifold of constant ϕ-sectional curvature c and let M be an (n+1)-dimensional anti-invariant submanifold of $\bar{M}$. In the following we compute the Laplacian of the square of the length of the second fundamental form of M.

By (2.5) of Chapter II we have

$$K^a_{ikj} = h^a_{ijk} - h^a_{ikj}.$$

Since M is anti-invariant in $\bar{M}$, (2.3) implies that $K^a_{ikj} = 0$. Thus we have the Codazzi equation

$$h^a_{ijk} - h^a_{ikj} = 0. \tag{3.1}$$

Therefore, substituting (2.3) into (3.18) of Chapter II, we find

$$\begin{aligned} \sum_{a,i,j} h^a_{ij}\Delta h^a_{ij} &= \sum_{a,i,j,k} h^a_{ij}h^a_{kkij} + \frac{1}{4}(c+3)(n+1)\sum_a \mathrm{Tr}A_a^2 \\ &- \frac{1}{4}(c-1)\sum_\lambda \mathrm{Tr}A_\lambda^2 - \frac{1}{2}(c+1)\sum_x (\mathrm{Tr}A_x)^2 - \frac{1}{4}(c+3)\sum_\lambda (\mathrm{Tr}A_\lambda)^2 \\ &- \frac{1}{2}(c-1)n(n+1) + \sum_{a,b}\{\mathrm{Tr}(A_aA_b - A_bA_a)^2 - [\mathrm{Tr}(A_aA_b)]^2 \\ &+ \mathrm{Tr}A_b\mathrm{Tr}(A_aA_bA_a)\}. \end{aligned} \tag{3.2}$$

Lemma 3.1. Let M be an (n+1)-dimensional anti-invariant submanifold of a Sasakian space form $\bar{M}^{2m+1}(c)$. Then we have

$$\begin{aligned} \sum_{a,i,j} h^a_{ij}\Delta h^a_{ij} &= \sum_{a,i,j,k} h^a_{ij}h^a_{kkij} + \frac{1}{4}(c+3)(n+1)\sum_a \mathrm{Tr}H_a^2 \\ &- \frac{1}{2}(c+3)\sum_x (\mathrm{Tr}H_x)^2 - \frac{1}{4}(c-1)\sum_\lambda \mathrm{Tr}H_\lambda^2 - 4\sum_\lambda \mathrm{Tr}H_\lambda^2 - \frac{1}{4}(c-1)\sum_\lambda (\mathrm{Tr}H_\lambda)^2 \\ &+ \sum_{a,b}\{\mathrm{Tr}(H_aH_b - H_bH_a)^2 - [\mathrm{Tr}(H_aH_b)]^2 + \mathrm{Tr}H_b\mathrm{Tr}(H_aH_bH_a)\}. \end{aligned} \tag{3.3}$$

Proof. First of all, by (1.3) we have

$$\text{(3.4)} \qquad \mathrm{Tr}A_x^2 = \mathrm{Tr}H_x^2 + 2, \qquad \mathrm{Tr}A_\lambda^2 = \mathrm{Tr}H^2, \qquad \mathrm{Tr}A_a = \mathrm{Tr}H_a.$$

From (1.2) we also have the equations:

$$\text{(3.5)} \qquad \sum_{a,b} \mathrm{Tr}(A_aA_b - A_bA_a)^2 = \sum_{a,b} \mathrm{Tr}(H_aH_b - H_bH_a)^2 - 4\sum_x (\mathrm{Tr}H_x)^2$$
$$- 4\sum_a \mathrm{Tr}H_a^2 + 8\sum_x \mathrm{Tr}H_x^2 - 2n(n-1),$$

$$\text{(3.6)} \qquad \sum_{a,b} [\mathrm{Tr}(A_aA_b)]^2 = \sum_a (\mathrm{Tr}A_a^2)^2 = \sum_a (\mathrm{Tr}H_a^2)^2 + 4\sum_x \mathrm{Tr}H_x^2 + 4n,$$

$$\text{(3.7)} \qquad \sum_{a,b} \mathrm{Tr}A_b\mathrm{Tr}(A_aA_bA_a) = \sum_{a,b} \mathrm{Tr}H_b\mathrm{Tr}(H_aH_bH_a) + 2\sum_x (\mathrm{Tr}H_x)^2 + \sum_a (\mathrm{Tr}H_a)^2.$$

Substituting (3.5), (3.6) and (3.7) into (3.2) and using (3.4), we have equation (3.3)

From (3.12) of Chapter II, (1.1) and (1.2) we have

$$\text{(3.8)} \qquad \sum_{a,i,j} h_{ij}^a \Delta h_{ij}^a = \frac{1}{2}\Delta S - \sum_{a,i,j,k} (h_{ijk}^a)^2$$
$$= \frac{1}{2}\Delta S - \sum_{a,x,y,z} (h_{xyz}^a)^2 - 3\sum_\lambda \mathrm{Tr}H_\lambda^2,$$

$$\text{(3.9)} \qquad \sum_{a,i,j,k} h_{ij}^a h_{kkij}^a = \sum_{a,x,j,k} h_{xj}^a h_{kkxj}^a - \sum_\lambda (\mathrm{Tr}H_\lambda)^2.$$

Substituting (3.8) and (3.9) into (3.3), we have

Lemma 3.2. Let M be an (n+1)-dimensional anti-invariant submanifold of a Sasakian space form $\bar{M}^{2m+1}(c)$. Then we have

$$\text{(3.10)} \qquad \frac{1}{2}\Delta S - \sum_{a,x,y,z} (h_{xyz}^a)^2 = \sum_{a,x,j,k} h_{xj}^a h_{kkxj}^a + \frac{1}{4}(c+3)\sum_a [n\mathrm{Tr}H_a^2$$
$$- (\mathrm{Tr}H_a)^2] + \frac{1}{4}(c+3)\sum_x [\mathrm{Tr}H_x^2 - (\mathrm{Tr}H_x)^2] + \sum_{a,b} \{\mathrm{Tr}(H_aH_b - H_bH_a)^2$$

$$- [\mathrm{Tr}(H_aH_b)]^2 + \mathrm{Tr}H_b\mathrm{Tr}(H_aH_bH_a)\}.$$

Moreover (3.8) and (3.9) imply the following lemmas.

Lemma 3.3. Let M be an (n+1)-dimensional anti-invariant submanifold of a (2m+1)-dimensional Sasakian manifold $\bar{M}$. If the second fundamental forms of M are parallel, then $H_\lambda = 0$ for all λ.

Lemma 3.4. Let M be an (n+1)-dimensional anti-invariant submanifold of a (2m+1)-dimensional Sasakian manifold $\bar{M}$. If the mean curvature vector of M is parallel, then we have $\mathrm{Tr}H_\lambda = 0$ for all λ.

From Lemmas 1.1 and 3.2 we obtain

Lemma 3.5. Let M be an (n+1)-dimensional anti-invariant submanifold of a Sasakian space form $\bar{M}^{2m+1}(c)$. If the f-structure in the normal bundle is parallel, then we have

$$\begin{aligned}(3.11)\qquad \frac{1}{2}\Delta S - \sum_{a,x,y,z}(h^a_{xyz})^2 &= \sum_{a,x,j,k} h^a_{xj}h^a_{kkxj} + \frac{1}{4}(c+3)(n+1)\sum_x \mathrm{Tr}H_x^2 \\ &- \frac{1}{2}(c+3)\sum_x(\mathrm{Tr}H_x)^2 + \sum_{x,y}\{\mathrm{Tr}(H_xH_y - H_yH_x)^2 - [\mathrm{Tr}(H_xH_y)]^2 \\ &+ \mathrm{Tr}H_y\mathrm{Tr}(H_xH_yH_x)\}.\end{aligned}$$

Remark. If the f-structure in the normal bundle is parallel, then (3.8) and (3.9) imply that

$$\sum_{a,i,j,k} h^a_{ij}h^a_{kkij} = \sum_{a,x,j,k} h^a_{xj}h^a_{kkxj},$$

$$\sum_{a,i,j} h^a_{ij}\Delta h^a_{ij} = \frac{1}{2}\Delta S - \sum_{a,x,y,z}(h^a_{xyz})^2,$$

$$\sum_{a,x,y,z}(h^a_{xyz})^2 = \sum_{a,i,j,k}(h^a_{ijk})^2.$$

Lemma 3.6. Let M be an (n+1)-dimensional anti-invariant minimal submanifold of a Sasakian space form $\bar{M}^{2m+1}(c)$. Then we have

$$(3.12)\qquad \frac{1}{2}\Delta S - \sum_{a,x,y,z}(h^a_{xyz})^2 = \frac{1}{4}(c+3)n\sum_a \mathrm{Tr}H_a^2 + \frac{1}{4}(c+3)\sum_x \mathrm{Tr}H_x^2$$

$$+ \sum_{a,b}\{\mathrm{Tr}(H_aH_b - H_bH_a)^2 - [\mathrm{Tr}(H_aH_b)]^2\}.$$

Lemma 3.7. Let M be an (n+1)-dimensional anti-invariant minimal submanifold of a Sasakian space form $\bar{M}^{2m+1}(c)$. If the f-structure in the normal bundle is parallel, then we have

$$(3.13)\qquad \frac{1}{2}\Delta S - \sum_{a,x,y,z}(h^a_{xyz})^2 = \frac{1}{4}(c+3)(n+1)\sum_x \mathrm{Tr}H_x^2$$

$$+ \sum_{x,y}\{\mathrm{Tr}(H_xH_y - H_yH_x)^2 - [\mathrm{Tr}(H_xH_y)]^2\}.$$

§4. The length of the second fundamental form

In the sequel, we put

$$T_{ab} = \sum_{x,y} h^a_{xy}h^b_{xy}, \qquad T_a = T_{aa}, \qquad T = \sum_a T_a.$$

We first prove

Theorem 4.1. Let M be an (n+1)-dimensional compact orientable anti-invariant minimal submanifold of a Sasakian space form $\bar{M}^{2m+1}(c)$. If the f-structure in the normal bundle is parallel, then we have

$$(4.1)\qquad \int_M [(2 - \frac{1}{n})T - \frac{1}{4}(c+3)(n+1)]T{*}1 \geq \int_M \sum_{a,i,j,k}(h^a_{ijk})^2{*}1 \geq 0.$$

Proof. We can rewrite (3.13) as follows:

$$\frac{1}{4}\Delta S - \sum_{a,i,j,k}(h^a_{ijk})^2 = \frac{1}{4}(c+3)(n+1)T - \sum_x T_x^2 + \sum_{x,y}\mathrm{Tr}(H_xH_y - H_yH_x)^2. \tag{4.2}$$

Applying Lemma 3.5 of Chapter III, we obtain

$$\begin{aligned} - \sum_{x,y}\mathrm{Tr}(H_xH_y - H_yH_x)^2 &+ \sum_x T_x^2 - \frac{1}{4}(c+3)(n+1)T \\ &\leq 2\sum_{x\neq y} T_xT_y + \sum_x T_x^2 - \frac{1}{4}(c+3)(n+1)T \\ &= [(2 - \frac{1}{n})T - \frac{1}{4}(c+3)(n+1)]T - \frac{1}{n}\sum_{x>y}(T_x - T_y)^2. \end{aligned} \tag{4.3}$$

Since M is compact and orientable, we have (4.1) by (4.2) and (4.3).

As a direct consequence of Theorem 4.1, we have

Corollary 4.1. Let M be an (n+1)-dimensional compact orientable anti-invariant minimal submanifold of a Sasakian space form $\bar{M}^{2m+1}(c)$. If the f-structure in the normal bundle is parallel, then either T = 0 or $T = (c+3)n(n+1)/4(2n-1)$ or at some point x of M, $T(x) > (c+3)n(n+1)/4(2n-1)$.

In the sequel, we study the case in which $T = (c+3)n(n+1)/4(2n-1)$. Now we assume that the ambient manifold $\bar{M}$ is of constant curvature 1, that is, c = 1, then the square of the length of the second fundamental form of M is given by $S = n(5n-1)/(2n-1)$.

Theorem 4.2. Let M be an (n+1)-dimensional anti-invariant minimal submanifold of a Sasakian space form $\bar{M}^{2m+1}(1)$. If the f-structure in the normal bundle is parallel and if $S = n(5n-1)/(2n-1)$, then n = 2 and M is a flat anti-invariant minimal submanifold of some $\bar{M}^5(1)$ in $\bar{M}^{2m+1}(1)$, where $\bar{M}^5(1)$ is a totally geodesic invariant submanifold of $\bar{M}^{2m+1}(1)$ of dimension 5. With respect to an adapted dual orthonormal frame field

$\eta=\omega^0,\omega^1,\omega^2,\omega^{1*},\omega^{2*}$ in $\bar{M}^5(1)$, the connection form (ω^A_B) of $\bar{M}^5(1)$, restricted to M, is given by

$$\begin{pmatrix} 0 & 0 & 0 & -\omega^1 & -\omega^2 \\ 0 & 0 & 0 & \omega^0+\lambda\omega^2 & \lambda\omega^1 \\ 0 & 0 & 0 & \lambda\omega^1 & \omega^0-\lambda\omega^2 \\ \omega^1 & \omega^0+\lambda\omega^2 & \lambda\omega^1 & 0 & 0 \\ \omega^2 & \lambda\omega^1 & \omega^0-\lambda\omega^2 & 0 & 0 \end{pmatrix}, \quad \lambda = \frac{1}{\sqrt{2}}.$$

Proof. From the assumption on the square of the length of the second fundamental form of M, we have $T = n(n+1)/(2n-1)$. Consequently (4.2) and (4.3) imply that the second fundamental form of M is parallel. Moreover, by Lemma 3.5 of Chapter III, we have

$$(4.4) \qquad \sum_{x,y} (T_x - T_y)^2 = 0,$$

$$(4.5) \qquad -\operatorname{Tr}(H_xH_y - H_yH_x)^2 = 2\operatorname{Tr}H_x^2\operatorname{Tr}H_y^2.$$

Since the f-structure in the normal bundle is parallel, we find that $h^\lambda_{ij} = 0$, i.e., $A_\lambda = H_\lambda = 0$. By (4.4) and (4.5) we can assume that $H_x = 0$ for $x = 3,\dots,n$ and $T_x = T_y$ for all x and y. Thus we must have $n = 2$. Thus we obtain

$$(4.6) \qquad H_1 = \lambda \begin{pmatrix} 0 & 1 \\ 1 & 0 \end{pmatrix}, \qquad H_2 = \lambda \begin{pmatrix} 1 & 0 \\ 0 & -1 \end{pmatrix},$$

by using $h^1_{12} = \lambda = h^2_{11}$, and consequently (1.3) and (4.6) imply

$$(4.7) \qquad A_1 = \begin{pmatrix} 0 & 1 & 0 \\ 1 & 0 & \lambda \\ 0 & \lambda & 0 \end{pmatrix}, \qquad A_2 = \begin{pmatrix} 0 & 0 & 1 \\ 0 & \lambda & 0 \\ 1 & 0 & -\lambda \end{pmatrix}.$$

But we have $h_{ij}^{\lambda} = 0$, and hence

$$(4.8) \qquad \omega_i^{\lambda} = 0.$$

Since the second fundamental form of M is parallel, using (3.10) of Chapter II, we have

$$(4.9) \qquad dh_{ij}^a = h_{ik}^a \omega_j^k + h_{kj}^a \omega_i^k - h_{ij}^b \omega_b^a.$$

From (4.9) we find $h_{ij}^x \omega_{x*}^{\lambda} = 0$, which implies that

$$(4.10) \qquad \omega_{x*}^{\lambda} = 0.$$

Putting a = 1, i = 1 and j = 0 in (4.9), we see that $d\lambda = \omega_0^2 = 0$, which means that λ is a constant. Since T = 2, we get $4\lambda^2 = 2$. Thus we may assume that $\lambda = 1/\sqrt{2}$ and obtain

$$(4.11) \qquad \omega_i^{x*} \neq 0.$$

Since we have (4.8), (4.10) and (4.11) we can define a distribution L defined by

$$\omega^{\lambda} = 0, \qquad \omega_i^{\lambda} = 0, \qquad \omega_{i*}^{\lambda} = 0.$$

It easily follows from the structure equations that

$$d\omega^{\lambda} = 0, \qquad d\omega_i^{\lambda} = 0, \qquad d\omega_{i*}^{\lambda} = 0.$$

Therefore L is a 5-dimensional completely integrable distribution. We consider the maximal integral submanifold $\bar{M}(x)$ of L through $x \in M$. Then $\bar{M}(x)$ is of dimension 5 and by construction it is totally geodesic invariant submanifold of $\bar{M}^{2m+1}(1)$ and hence it is of constant curvature 1. Moreover M is an anti-invariant submanifold of $\bar{M}^5(1)$ and minimal in $\bar{M}^5(1)$. In $\bar{M}^5(1)$, (4.7) and (4.9) imply the equations:

$$
\begin{aligned}
&\omega_1^0 = \omega^{1*} = 0, && \omega_2^0 = \omega^{2*} = 0,\\
&\omega_{1*}^0 = -\omega^1, && \omega_{2*}^0 = -\omega^2,\\
&\omega_1^{2*} = \lambda\omega^1, && \omega_1^{1*} = \omega^0 + \lambda\omega_1^2,\\
&\omega_2^{2*} = \omega^0 - \lambda\omega^2, && \omega_1^2 = 0.
\end{aligned}
\tag{5.12}
$$

From the Gauss equation (3.7) of Chapter II and (4.7) we easily see that M is flat. These considerations prove our assertion.

§5. Anti-invariant minimal submanifolds of S^5

We consider the following example of anti-invariant submanifolds.

Example 5.1. Let S^5 denote a 5-dimensional sphere in C^3 with standard Sasakian structure. Let J be the almost complex structure of C^3 given by

$$
J = \begin{pmatrix}
0 & -1 & & & & \\
1 & 0 & & & & \\
 & & 0 & -1 & & \\
 & & 1 & 0 & & \\
 & & & & 0 & -1 \\
 & & & & 1 & 0
\end{pmatrix} .
$$

Let $S^1(\frac{1}{\sqrt{3}}) = \{z \in C : |z|^2 = \frac{1}{3}\}$ be a circle of radius $1/\sqrt{3}$ and consider

$$M^3 = S^1(\frac{1}{\sqrt{3}}) \times S^1(\frac{1}{\sqrt{3}}) \times S^1(\frac{1}{\sqrt{3}})$$

in S^5 in C^3, which is obviously flat. The position vector X of M^3 in S^5 in C^3 has components given by

$$X = \frac{1}{\sqrt{3}} \begin{pmatrix} \cos u^1 \\ \sin u^1 \\ \cos u^2 \\ \sin u^2 \\ \cos u^3 \\ \sin u^3 \end{pmatrix},$$

u^1, u^2 and u^3 being parameters on each $S^1(\frac{1}{\sqrt{3}})$. Putting $X_i = \partial X/\partial u^i$, we find

$$X_1 = \frac{1}{\sqrt{3}} \begin{pmatrix} -\sin u^1 \\ \cos u^1 \\ 0 \\ 0 \\ 0 \\ 0 \end{pmatrix}, \quad X_2 = \frac{1}{\sqrt{3}} \begin{pmatrix} 0 \\ 0 \\ -\sin u^2 \\ \cos u^2 \\ 0 \\ 0 \end{pmatrix}, \quad X_3 = \frac{1}{\sqrt{3}} \begin{pmatrix} 0 \\ 0 \\ 0 \\ 0 \\ -\sin u^3 \\ \cos u^3 \end{pmatrix}.$$

The structure vector ξ on S^5 is given by

$$\xi = JX = \frac{1}{\sqrt{3}} \begin{pmatrix} -\sin u^1 \\ \cos u^1 \\ -\sin u^2 \\ \cos u^2 \\ -\sin u^3 \\ \cos u^3 \end{pmatrix}.$$

Thus $\xi = X_1 + X_2 + X_3$, and consequently the structure vector ξ is tangent to M^3. On the other hand, the structure tensors (ϕ,ξ,η) of S^5 satisfy

$$\phi X_i = JX_i + \eta(X_i)X, \qquad i = 1,2,3,$$

which shows that ϕX_i is normal to M^3 for all i. Therefore M^3 is an anti-invariant submanifold of S^5. Moreover M^3 is a minimal submanifold of S^5 with $S = 6$ and the normal connection of M^3 is flat (see also Proposition 2.2). M^3 is also an anti-invariant submanifold of S^{2m+1} $(m > 2)$ with parallel f-structure in the normal bundle.

From Theorem 4.1 and Example 5.1, we have

Theorem 5.1. Let M be an (n+1)-dimensional compact orientable anti-invariant minimal submanifold of S^{2m+1}. If the f-structure in the normal bundle is parallel and if $S = (5n^2-n)/(2n-1)$, then

$$M = S^1(\frac{1}{\sqrt{3}}) \times S^1(\frac{1}{\sqrt{3}}) \times S^1(\frac{1}{\sqrt{3}}) \quad \text{in an } S^5 \text{ in } S^{2m+1}.$$

Theorem 5.2 (Yano-Kon [79]). Let M be an (n+1)-dimensional compact orientable anti-invariant minimal submanifold of S^{2n+1}. If $S = (5n^2-n)/(2n-1)$, then

$$M = S^1(\frac{1}{\sqrt{3}}) \times S^1(\frac{1}{\sqrt{3}}) \times S^1(\frac{1}{\sqrt{3}}).$$

When the second fundamental form of M is parallel, by Lemma 3.3 and by a method quite similar to that used in proof of Theorem 5.1, we have the following

Theorem 5.3. Let M be an (n+1)-dimensional compact orientable anti-invariant minimal submanifold of S^{2m+1}. If the second fundamental form of M is parallel and if $S = (5n^2-n)/(2n-1)$, then

$$M = S^1(\frac{1}{\sqrt{3}}) \times S^1(\frac{1}{\sqrt{3}}) \times S^1(\frac{1}{\sqrt{3}}) \quad \text{in an } S^5 \text{ in } S^{2m+1}.$$

§6. Flat normal connection

Proposition 6.1. Let M be an (n+1)-dimensional anti-invariant minimal submanifold with flat normal connection of S^{2m+1}. If the f-structure in the normal bundle is parallel and if $S = n+1$, then $n = 1$ and M is flat.

Proof. From the assumption we have $A_\lambda = 0$ for all λ and $A_xA_y = A_yA_x$ for all x and y. Thus (3.2) becomes

$$(6.1) \qquad \sum_{a,i,j} h^a_{ij}\Delta h^a_{ij} = (n+1)\sum_x \mathrm{Tr}A_x^2 - \sum_{x,y}[\mathrm{Tr}(A_xA_y)]^2.$$

Putting

$$S_{ab} = \sum_{i,j} h^a_{ij}h^b_{ij}, \qquad S_{aa} = S_a, \qquad S = \sum_a S_a,$$

we can rewrite equation (6.1) as

$$(6.2) \qquad \sum_{a,i,j} h^a_{ij}\Delta h^a_{ij} = (n+1)S - \sum_x S_x^2 = (n+1)S - (\sum_x S_x)^2 + \sum_{x\neq y} S_xS_y.$$

Since S = constant, we have $\sum_{a,i,j} h^a_{ij}\Delta h^a_{ij} = -\sum_{a,i,j,k}(h^a_{ijk})^2$. From this and (6.2) we have

$$(6.3) \qquad 0 \leq \sum_{a,i,j,k}(h^a_{ijk})^2 = [S-(n+1)]S - \sum_{x\neq y} S_xS_y \leq [S-(n+1)]S.$$

We see from (6.3) that $S = n+1$, then the second fundamental form of M is parallel and $\sum_{x\neq y} S_xS_y = 0$. Thus we can assume that $S_1 = n+1$ and $S_x = 0$ for $x = 2,\dots,n$ and obtain

$$(6.4) \qquad \sum_{i,j} (h^1_{ij})^2 = n + 1, \qquad h^x_{ij} = 0 \qquad \text{for all } x > 1 \text{ and } i,j.$$

Using (1.3) we see that $\sum_{i,j} (h^1_{ij})^2 = 2 + \sum_{x,y} (h^1_{xy})^2$. Since $h^1_{xy} = h^x_{1y} = h^y_{1x}$ $= 0$ unless $x = y = 1$, we have

$$(6.5) \qquad \sum_{i,j} (h^1_{ij})^2 = 2 + (h^1_{11})^2.$$

If M is minimal, the second fundamental form of M satisfies that $\sum_x (h^1_{xx}) = 0$, which implies $h^1_{11} = 0$. Consequently, we must have $n+1 = 2$, that is, $n = 1$. Moreover, Proposition 2.2 shows that M is flat.

Example 6.1. Let $S^1(\frac{1}{\sqrt{2}})$ be a circle of radius $1/\sqrt{2}$. Then $S^1(\frac{1}{\sqrt{2}}) \times S^1(\frac{1}{\sqrt{2}})$ is an anti-invariant minimal submanifold of S^3 and also is an anti-invariant minimal submanifold of S^{2m+1} $(m > 1)$ with parallel f-structure in the normal bundle. In this case, we have $S = 2$.

By a well-known theorem of Chern-do Carmo-Kobayashi [10] and Proposition 6.1 we have

Theorem 6.1. Let M be an (n+1)-dimensional compact orientable anti-invariant minimal submanifold of S^{2m+1}. If the normal connection of M is flat, the f-structure in the normal bundle is parallel and $S = n+1$, then

$$M = S^1(\frac{1}{\sqrt{2}}) \times S^1(\frac{1}{\sqrt{2}}) \qquad \text{in an } S^3 \text{ in } S^{2m+1}.$$

Theorem 6.2 (Yano-Kon [79]). Let M be an (n+1)-dimensional compact orientable anti-invariant minimal submanifold with flat normal connection of S^{2n+1}. If $S = n+1$, then

$$M = S^1(\frac{1}{\sqrt{2}}) \times S^1(\frac{1}{\sqrt{2}}) \qquad \text{in } S^3.$$

Example 6.2. Let J be the almost complex structure of C^{n+1} given by

$$J = \begin{pmatrix} \begin{array}{cc|} 0 & -1 \\ 1 & 0 \\ \hline \end{array} & & & \\ & \ddots & & \\ & & \ddots & \\ & & & \begin{array}{|cc} \hline 0 & -1 \\ 1 & 0 \end{array} \end{pmatrix}.$$

Let $S^1(r_i) = \{z_i \in C : |z_i|^2 = r_i^2\}$, $i = 1,\ldots,n+1$. We consider

$$M^{n+1} = S^1(r_1) \times \ldots\ldots \times S^1(r_{n+1})$$

in C^{n+1} such that $r_1^2 + \ldots\ldots + r_{n+1}^2 = 1$. Then M^{n+1} is a flat submanifold of S^{2n+1} with parallel mean curvature vector and with flat normal connection (for instance, Erbacher [12] and Yano-Ishihara [76]).

The position vector X of M^{n+1} in C^{n+1} has components given by

$$X = \begin{pmatrix} r_1 \cos u^1 \\ r_1 \sin u^1 \\ \cdot \\ \cdot \\ \cdot \\ r_{n+1} \cos u^{n+1} \\ r_{n+1} \sin u^{n+1} \end{pmatrix}, \qquad r_1^2 + \ldots\ldots + r_{n+1}^2 = 1,$$

and is an outward unit normal vector of S^{2n+1} in C^{n+1}. Putting $X_i = \partial_i X = \partial X/\partial u^i$, we have

$$X_1 = r_1 \begin{pmatrix} -\sin u^1 \\ \cos u^1 \\ 0 \\ \cdot \\ \cdot \\ \cdot \\ 0 \end{pmatrix}, \ldots\ldots\ldots\ldots, X_{n+1} = r_{n+1} \begin{pmatrix} 0 \\ \cdot \\ \cdot \\ \cdot \\ 0 \\ -\sin u^{n+1} \\ \cos u^{n+1} \end{pmatrix}.$$

The vector field ξ on S^{2n+1} is given by

$$\xi = JX = \begin{pmatrix} -r_1 \sin u^1 \\ r_1 \cos u^1 \\ \cdot \\ \cdot \\ \cdot \\ -r_{n+1} \sin u^{n+1} \\ r_{n+1} \cos u^{n+1} \end{pmatrix},$$

and hence we have $\xi = X_1 + \ldots\ldots + X_{n+1}$, which means that the structure vector field ξ is tangent to M^{n+1}. We see that the structure tensors (ϕ,ξ,η) of S^{2n+1} satisfy

$$\phi X_i = JX_i + \eta(X_i)X, \qquad i = 1,\ldots\ldots,n+1,$$

and hence ϕX_i is normal to M for all i. Therefore M^{n+1} is an anti-invariant submanifold of S^{2n+1}. Moreover, M^{n+1} is an anti-invariant submanifold of S^{2m+1} $(m > n)$ with parallel f-structure in the normal bundle, with parallel mean curvature vector and with flat normal connection.

Theorem 6.3. Let M be an (n+1)-dimensional compact orientable anti-invariant submanifold with parallel mean curvature vector and with flat normal connection of S^{2m+1}. If the f-structure in the normal bundle is parallel, then

$$M = S^1(r_1) \times \cdots \times S^1(r_{n+1}) \quad \text{in an } S^{2n+1} \text{ in } S^{2m+1},$$

where $r_1^2 + \cdots + r_{n+1}^2 = 1$.

Proof. If M is regarded as a submanifold in C^{m+1}, then M is an anti-invariant submanifold of C^{m+1}. Indeed, for any vector field X of M, we have $\phi X = JX + \eta(X)N$ where N is the outward unit normal vector of S^{2m+1} in C^{m+1} and hence ϕX is normal to M if and only if JX is normal to M. Since the mean curvature vector of M in S^{2m+1} is parallel, we see that the mean curvature vector of M in C^{m+1} is also parallel with respect to the normal connection of M in C^{m+1}. Moreover, we see that the normal connection of M in C^{m+1} is flat and the f-structure in the normal bundle of M in C^{m+1} is parallel. Therefore, using Theorem 7.1 of Chapter III, we have our assertion.

When the second fundamental form of M is parallel, by Lemma 3.3, we have $A_\lambda = 0$ for all λ. Thus we have

Theorem 6.4. Let M be an (n+1)-dimensional compact orientable anti-invariant submanifold with parallel second fundamental form of S^{2m+1}. If the normal connection of M is flat, then

$$M = S^1(r_1) \times \cdots \times S^1(r_{n+1}) \quad \text{in an } S^{2n+1} \text{ in } S^{2m+1},$$

where $r_1^2 + \cdots + r_{n+1}^2 = 1$.

Theorem 6.5 (Yano-Kon [79]). Let M be an (n+1)-dimensional compact orientable anti-invariant submanifold with parallel mean curvature vector of S^{2n+1}. If the normal connection of M is flat, then we have

$$M = S^1(r_1) \times \cdots \times S^1(r_{n+1}),$$

where $r_1^2 + \ldots\ldots + r_{n+1}^2 = 1$.

§7. Pinching theorems

First of all, we prove

Theorem 7.1. Let M be an (n+1)-dimensional compact orientable anti-invariant minimal submanifold of a Sasakian space form $\bar{M}^{2m+1}(c)$ $(c > -3)$. If $T \leq \frac{1}{4}(c+3)nq/(2q-1)$, where $q = 2m-n$, then $T = 0$.

Proof. From (3.12) and Lemma 3.5 of Chapter III we have

$$(7.1) \qquad \sum_{a,x,y,z} (h^a_{xyz})^2 - \frac{1}{2}\Delta S = -\sum_{a,b} Tr(H_aH_b - H_bH_a)^2 + \sum_a T_a^2$$

$$- \frac{1}{4}(c+3)nT - \frac{1}{4}(c+3)\sum_x TrH_x^2$$

$$\leq [(2 - \frac{1}{q})T - \frac{1}{4}(c+3)n]T - \frac{1}{q}\sum_{a>b}(T_a - T_b)^2 - \frac{1}{4}(c+3)\sum_x T_x.$$

Since M is compact and orientable, we have

$$\int_M \sum_{a,x,y,z} (h^a_{xyz})^2 *1 \leq \int_M [(2 - \frac{1}{q})T - \frac{1}{4}(c+3)n]T*1.$$

Therefore, by the assumption and Lemma 3.5 of Chapter III, we have

$$(7.2) \qquad \sum_{a\,b}(T_a - T_b)^2 = 0, \qquad \sum_x T_x = 0.$$

Equations (7.2) imply that $T_a = T_b$ for all a and b and $T_x = 0$ for all x. Therefore we have $T_a = 0$ for all a and hence $T = 0$.

From Lemmas 3.3 and 3.6, using (4.3), we have

Theorem 7.2. Let M be an (n+1)-dimensional compact orientable anti-invariant minimal submanifold with parallel second fundamental form of a Sasakian space form $\bar{M}^{2m+1}(c)$. If $T < (c+3)n(n+1)/4(2n-1)$, then $T = 0$.

In a latter section, we shall give an example of anti-invariant submanifold with $T = 0$. When $T = 0$, by Proposition 7.2 of Chapter II, M is locally a Riemannian direct product $M^n \times M^1$, where M^n is a totally geodesic anti-invariant submanifold of $\bar{M}$ normal to the structure vector field and M^1 is generated by ξ.

§8. The condition $H_aH_b = H_bH_a$

First of all, we prepare

Lemma 8.1. Let M be an (n+1)-dimensional anti-invariant submanifold of a (2m+1)-dimensional Sasakian manifold $\bar{M}$. If $H_aH_b = H_bH_a$ for all a and b, then we can choose an orthonormal frame for which H_x is of the form

$$H_x = \begin{pmatrix} 0 & & & & & & \\ & \ddots & & & & & \\ & & 0 & & & & \\ & & & \lambda_x & & & \\ & & & & 0 & & \\ & & & & & \ddots & \\ & & & & & & 0 \end{pmatrix} x,$$

where $x = 1,\dots,n$, that is, $h^x_{yz} = 0$ unless $x = y = z$.

Proof. If $H_aH_b = H_bH_a$, we can choose an orthonormal frame $e_1,\dots,e_n$ for which all H_a are simultaneously diagonal, i.e., $h^a_{xy} = 0$ when $x \neq y$, that is, $h^z_{xy} = 0$ when $x \neq y$. Thus (1.2) shows that $h^z_{xy} = 0$ unless $x = y = z$.

Corollary 8.1. Let M be an (n+1)-dimensional anti-invariant minimal submanifold of a (2m+1)-dimensional Sasakian manifold $\bar{M}$. If the f-structure in the normal bundle is parallel and if $H_aH_b = H_bH_a$ for all a and b, then $H_a = 0$ for all a.

Proof. From (1.3) we see that M is minimal if and only if $TrH_a = 0$ for all a. Thus Lemma 8.1 implies that $\lambda_x = 0$, that is, $H_x = 0$. Since the f-structure in the normal bundle is parallel, we also have $H_\lambda = 0$. Therefore $H_a = 0$ for all a.

In the following, if $h^a_{xy} = \delta_{xy}(TrH_a)/n$ for all a, x and y, then we say that M is H-totally umbilical. Then we have

Corollary 8.2. Let M be an (n+1)-dimensional $(n > 1)$ anti-invariant submanifold of a (2m+1)-dimensional Sasakian manifold. If the f-structure in the normal bundle is parallel and if M is H-totally umbilical, then $H_a = 0$ for all a.

Proof. Since $h^a_{xy} = \delta_{xy}(TrH_a)/n$, we have $H_aH_b = H_bH_a$. By Lemma 8.1 we have $h^x_{yz} = 0$ unless $x = y = z$. On the other hand, we have $h^x_{yz} = \lambda_x\delta_{yz}/n$. Putting here $y = z \neq x$, we have $\lambda_x = 0$, i.e., $H_x = 0$. Since the f-structure in the normal bundle is parallel, we have $H_a = 0$ for all a.

From (1.2) and (3.7) of Chapter II, we have

Lemma 8.2. Let $\bar{M}$ be a (2m+1)-dimensional Sasakian manifold and let M be an (n+1)-dimensional anti-invariant submanifold of $\bar{M}$ with parallel f-structure in the normal bundle. If $H_aH_b = H_bH_a$ for all a and b, then we have

$$R^x_{yzv} = K^x_{yzv}. \tag{8.1}$$

Proposition 8.1. Let M be an (n+1)-dimensional anti-invariant submanifold of a Sasakian space form $\bar{M}^{2m+1}(c)$ with parallel f-structure in the normal bundle. If $H_aH_b = H_bH_a$ for all a and b, then M is flat if and only if $c = -3$.

Proof. From the assumption and (8.1) we have

$$(8.2) \qquad R^x_{yzv} = \frac{1}{4}(c+3)(\delta_{xz}\delta_{yv} - \delta_{xv}\delta_{yz}).$$

From (8.2), combined with Proposition 7.2 of Chapter II, we have our assertion.

Lemma 8.3. Let M be an (n+1)-dimensional anti-invariant submanifold with parallel mean curvature vector and with parallel f-structure in the normal bundle of a Sasakian space form $\bar{M}^{2m+1}(c)$. If $H_aH_b = H_bH_a$ for all a and b, then we have

$$(8.3) \qquad \sum_{a,i,j,k}(h^a_{ijk})^2 = -\frac{1}{4}(c+3)(n-1)\sum_x \mathrm{Tr}H_x^2.$$

Proof. From the assumption and the Gauss equation (3.7) of Chapter II we see that the scalar curvature r of M is given by

$$(8.4) \qquad r = \frac{1}{4}n(n-1)(c+3) + \sum_x(\mathrm{Tr}H_x)^2 - \sum_x \mathrm{Tr}H_x^2.$$

Since $H_aH_b = H_bH_a$, we have $r = \frac{1}{4}n(n-1)(c+3)$. Since the mean curvature vector of M is parallel, by (8.4) we see that $\sum_x \mathrm{Tr}H_x^2$ is a constant and hence S is also a constant. By (3.11) we have

$$(8.5) \qquad \sum_{a,i,j,k}(h^a_{ijk})^2 = \frac{1}{4}(c+3)(n+1)\sum_x \mathrm{Tr}H_x^2 + \frac{1}{2}(c+3)\sum_x(\mathrm{Tr}H_x)^2$$
$$- \sum_{x,y}\{\mathrm{Tr}(H_xH_y - H_yH_x)^2 - [\mathrm{Tr}(H_xH_y)]^2 + \mathrm{Tr}H_y\mathrm{Tr}(H_xH_yH_x)\}.$$

By the assumption and Lemma 8.1 we have equation (8.3).

Proposition 8.2. Let M be an (n+1)-dimensional ($n > 1$) anti-invariant submanifold with parallel mean curvature vector and with parallel f-structure in the normal bundle of a Sasakian space form $\bar{M}^{2m+1}(c)$. If $H_a H_b = H_b H_a$, then either $H_a = 0$ for all a or $c \leq -3$.

Proposition 8.3. Let M be an (n+1)-dimensional ($n > 1$) anti-invariant submanifold with parallel second fundamental form of a Sasakian space form $\bar{M}^{2m+1}(c)$. If $H_a H_b = H_b H_a$, then either $H_a = 0$ for all a or $c = -3$ and M is flat.

Proof. Since the second fundamental form of M is parallel, by Lemma 3.3, we have $H_\lambda = 0$. Then (3.10) can be rewritten in the form (8.5). Since $H_a H_b = H_b H_a$, we obtain (8.3) and hence

$$\frac{1}{4}(c+3)(n-1)T = 0,$$

from which we have our assertion.

Proposition 8.4. Let M be an (n+1)-dimensional anti-invariant submanifold with parallel mean curvature vector and with parallel f-structure in the normal bundle of a Sasakian space form $\bar{M}^{2m+1}(c)$. If $H_a H_b = H_b H_a$ and $c \geq -3$, then the second fundamental form of M is parallel.

Example 8.1. Let E^{2n+1} be an Euclidean space with cartesian coordinates $(x^1,\ldots,x^n,y^1,\ldots,y^n,z)$. As in example of section 7 of Chapter I we derive the natural Sasakian structure in E^{2n+1} with constant ϕ-sectional curvature -3 and denote it by $E^{2n+1}(-3)$. Then we can consider the following natural imbedding of E^{n+1} into $E^{2n+1}(-3)$:

$$(x^1,\ldots,x^n,z) \longmapsto (x^1,\ldots,x^n,0,\ldots,0,z).$$

Then we easily see that E^{n+1} is an anti-invariant submanifold of $E^{2n+1}(-3)$ and the second fundamental forms of E^{n+1} in $E^{2n+1}(-3)$ are given by

$$
A_x = x\left(\begin{array}{c|c}
0 & 0\cdots 0\overset{x}{1}0\cdots 0 \\ \hline
0 & \\
\vdots & \\
0 & \\
1 & 0 \\
0 & \\
\vdots & \\
0 &
\end{array}\right), \quad \text{i.e.,} \quad H_x = 0 \text{ for all } x.
$$

Obviously $E^{2n+1}(-3)$ is a totally geodesic invariant submanifold of $E^{2m+1}(-3)$ $(m > n)$ and hence E^{n+1} is an anti-invariant submanifold of $E^{2m+1}(-3)$ with parallel f-structure in the normal bundle.

Example 8.2. Let $E^{2n+1}(-3)$ be as in Example 8.1. Let $S^1(r_i) = \{z_i \in C : |z_i|^2 = r_i^2\}$ be a circle of radius r_i, $i = 1,\ldots,n$. Now we consider

$$M^{n+1} = S^1(r_1) \times \cdots\cdots \times S^1(r_n) \times E^1.$$

Then we can define an imbedding of M^{n+1} into $E^{2n+1}(-3)$ by

$$(r_1 \cos u^1, \ldots\ldots, r_n \cos u^n, r_1 \sin u^1, \ldots\ldots, r_n \sin u^n, z).$$

In this case M^{n+1} is an anti-invariant submanifold of $E^{2n+1}(-3)$. Moreover the second fundamental forms of M^{n+1} are given by

$$
(8.6) \qquad A_x = x\left(\begin{array}{c|ccccccc}
0 & 0 & \cdots & 0 & \overset{x}{1} & 0 & \cdots & 0 \\ \hline
0 & 0 & & & & & & \\
\vdots & & \ddots & & & & & \\
0 & & & 0 & & & & \\
1 & & & & \lambda_x & & & \\
0 & & & & & 0 & & \\
\vdots & & & & & & \ddots & \\
0 & & & & & & & 0
\end{array}\right), \quad \text{i.e.,} \quad H_x = \begin{pmatrix}
0 & & & & & & \\
& \ddots & & & & & \\
& & 0 & & & & \\
& & & \lambda_x & & & \\
& & & & 0 & & \\
& & & & & \ddots & \\
& & & & & & 0
\end{pmatrix} x
$$

where $x = 1,\ldots,n$.

On the other hand, we consider

$$M^{n+1} = S^1(r_1) \times \ldots \times S^1(r_p) \times E^{n-p} \times E^1, \quad 1 \leq p \leq n.$$

This is an anti-invariant submanifold of $E^{2n+1}(-3)$ and the second fundamantal forms of M^{n+1} are given by (8.6) for $x = 1,\ldots,p$ and $H_x = 0$ for $x = p+1,\ldots,n$. Obviously, the second fundamental forms of M are parallel and $H_aH_b = H_bH_a$. Moreover M^{n+1} is an anti-invariant submanifold of E^{2m+1} (-3) $(m > n)$ with parallel f-structure in the normal bundle.

§9. Parallel mean curvature vector

In this section we prove the following theorems.

Theorem 9.1. Let M be a complete $(n+1)$-dimensional $(n > 1)$ anti-invariant submanifold of $E^{2m+1}(-3)$ with parallel mean curvature vector and with parallel f-structure in the normal bundle. If $H_aH_b = H_bH_a$, then M is E^{n+1} in an $E^{2n+1}(-3)$ in $E^{2m+1}(-3)$ or M is a product of the form

$$S^1(r_1) \times \ldots\ldots \times S^1(r_p) \times E^{n-p} \times E^1$$

in an $E^{2n+1}(-3)$ in $E^{2m+1}(-3)$, where $1 \leq p \leq n$.

Proof. From Proposition 8.4 we see that the second fundamental forms of M are parallel. Thus Proposition 8.3 implies that $H_a = 0$ for all a or M is flat. From these and Example 8.2 we have our assertion.

Theorem 9.2. Let M be a complete $(n+1)$-dimensional $(n > 1)$ anti-invariant submanifold with parallel second fundamental forms of a simply connected complete Sasakian space form $\bar{M}^{2m+1}(c)$. If $H_aH_b = H_bH_a$, and if $H_a \neq 0$ for some a, then M is a product of the form

$$S^1(r_1) \times \ldots\ldots \times S^1(r_p) \times E^{n-p} \times E^1$$

in an $E^{2n+1}(-3)$ in $E^{2m+1}(-3)$, where $1 \leq p \leq n$.

Proof. Since the second fundamental forms of M are parallel, we have $H_\lambda = 0$. By Proposition 8.3 we have $c = -3$ and hence Theorem 9.2 follows from Theorem 9.1.

Theorem 9.3. Let M be a complete (n+1)-dimensional ($n > 1$) anti-invariant submanifold of E^{2n+1} with parallel mean curvature vector and parallel f-structure in the normal bundle. If $H_a H_b = H_b H_a$, then M is E^{n+1} or M is a product of the form

$$S^1(r_1) \times \dots\dots \times S^1(r_p) \times E^{n-p} \times E^1, \quad 1 \leq p \leq n.$$

§10. Parallel second fundamental forms

Let M be an (n+1)-dimensional anti-invariant minimal submanifold with parallel second fundamental forms of a Sasakian space form $\bar{M}^{2m+1}(c)$. Then, by Lemma 3.3, we have $H_\lambda = 0$ for all λ. On the other hand, by Proposition 7.2 of Chapter II, we have

$$R^0_{ijk} = R^i_{0jk} = R^i_{j0k} = R^i_{jk0} = 0. \tag{10.1}$$

When $H_\lambda = 0$ for all λ, the Gauss equation (3.7) of Chapter II implies

$$\begin{aligned} R^x_{yzv} &= K^x_{yzv} + \sum_w (h^w_{xz}h^w_{yv} - h^w_{xv}h^w_{yz}) \\ &= \frac{1}{4}(c+3)(\delta_{xz}\delta_{yv} - \delta_{xv}\delta_{yz}) + \sum_w (h^w_{xz}h^w_{yv} - h^w_{xv}h^w_{yz}), \end{aligned} \tag{10.2}$$

from which, we find that the Ricci tensor R_{xy} of M is given by

$$R_{xy} = \frac{1}{4}(c+3)(n-1)\delta_{xy} - \sum_{w,z} h^w_{xz}h^w_{yz}. \tag{10.3}$$

The scalar curvature r of M is given by

$$r = \frac{1}{4}n(n-1)(c+3) - T. \tag{10.4}$$

From (10.3) and (10.4) we have

$$(10.5)\qquad \sum_{x,y} (R_{xy})^2 = \frac{1}{n}r^2 - \frac{1}{n}T^2 + \sum_{x,y} \mathrm{Tr}H_x^2H_y^2.$$

On the other hand, (1.2), (10.2) and (10.4) imply

$$(10.6)\qquad \sum_{x,y,z,v} (R^x_{yzv})^2 = \frac{2}{n(n-1)}r^2 - \frac{2}{n(n-1)}T^2 - \sum_{x,y} \mathrm{Tr}(H_xH_y - H_yH_x)^2.$$

Since the second fundamental forms of M are parallel, equation (3.12) becomes

$$(10.7)\qquad \frac{1}{4}(c+3)(n+1)T + \sum_{x,y} [\mathrm{Tr}(H_xH_y - H_yH_x)^2 - \mathrm{Tr}H_x^2H_y^2] = 0,$$

where we have used the fact that $\sum_{x,y} [\mathrm{Tr}(H_xH_y)]^2 = \sum_{x,y} \mathrm{Tr}H_x^2H_y^2$. Hence

Proposition 10.1. Let M be an (n+1)-dimensional anti-invariant minimal submanifold with parallel second fundamental forms of a Sasakian space form $\bar{M}^{2m+1}(c)$. If $c \leq -3$, then $T = 0$.

From (10.5), (10.6) and (10.7) we have

$$(10.8)\qquad \frac{(n+1)}{n(n-1)}Tr = \sum_{x,y,z,v} (R^x_{yzv})^2 - \frac{2}{n(n-1)}r^2 + \sum_{x,y} (R_{xy})^2 - \frac{1}{n}r^2.$$

We can easily verify that that the right hand side of (10.8) is non-negative. If the right side of (10.8) vanishes, then we have

$$(10.9)\qquad R^x_{yzv} = a(\delta_{xz}\delta_{yv} - \delta_{zv}\delta_{yz})$$

for some constant a.

Theorem 10.1. Let M be an (n+1)-dimensional anti-invariant minimal submanifold with parallel second fundamental forms of a Sasakian space form

$\bar{M}^{2m+1}(c)$. Then either $T = 0$ or M has the non-negative scalar curvature $r \geq 0$. Moreover if $r = 0$, then M is flat.

Proof. If $r < 0$, then (10.8) implies that $T = 0$. Except this case, we have $r \geq 0$. If $r = 0$, then (10.8) and (10.9) imply that M is flat.

Proposition 10.2. Let M be an $(n+1)$-dimensional anti-invariant minimal submanifold with parallel second fundamental forms of a Sasakian space form $\bar{M}^{2m+1}(c)$. Then $S \leq \frac{1}{4}(c+3)n(n+1)$. If the equality holds here, then M is locally a Riemannian direct product $M^n \times R^1$, where M^n is a hypersurface of M of constant curvature $\frac{1}{4}(c-1)$ and is totally geodesic in M.

Proof. From the assumption and (3.2) we have

$$(10.10) \qquad 0 \leq -\sum_{x,y} \mathrm{Tr}(A_xA_y - A_yA_x)^2 = \frac{1}{4}(c+3)(n+1)S - \sum_x S_x^2$$

$$= \frac{1}{4}(c+3)(n+1)S - \frac{1}{n}S^2 - \frac{1}{n}\sum_{x>y}(S_x - S_y)^2$$

$$\leq \frac{1}{n}[\frac{1}{4}(c+3)n(n+1) - S]S.$$

Since $S > 0$, we must have $S \leq \frac{1}{4}(c+3)n(n+1)$. If the equality holds here, then we have

$$(10.11) \qquad A_xA_y = A_yA_x, \qquad S_x = S_y \quad \text{for all } x \text{ and } y.$$

Thus the second fundamental forms of M are commutative. From this and Theorem 2.1 we have our assertion.

Proposition 10.3. Let M be an $(n+1)$-dimensional anti-invariant minimal submanifold with parallel second fundamental forms of a Sasakian space form $\bar{M}^{2m+1}(c)$. If $S = n(n+1)$, then M is locally a Riemannian direct product $M^n \times R^1$, where M^n is a flat hypersurface of M.

If we take S^{2m+1} as an ambient manifold, then (10.11) shows that the normal connection of M is flat. Therefore Theorem 6.4 implies the following

Theorem 10.2. Let M be an (n+1)-dimensional compact orientable anti-invariant minimal submanifold with parallel second fundamental forms of S^{2m+1}. If S = n(n+1), then

$$M = S^1\left(\frac{1}{\sqrt{n+1}}\right) \times \cdots\cdots \times S^1\left(\frac{1}{\sqrt{n+1}}\right) \quad \text{in an } S^{2n+1} \text{ in } S^{2m+1}.$$

§11. Surfaces with parallel mean curvature vector

Let M be an anti-invariant surface of a Sasakian space form $\bar{M}^{2m+1}(c)$. From (1.3) the second fundamental forms of M are given by

$$(11.1) \qquad A_1 = \begin{pmatrix} 0 & 1 \\ & \\ 1 & h^1_{11} \end{pmatrix}, \qquad A_\lambda = \begin{pmatrix} 0 & 0 \\ & \\ 0 & h^\lambda_{11} \end{pmatrix}.$$

If we assume that the mean curvature vector of M is parallel, then Lemma 3.4 implies that $\mathrm{Tr}A_\lambda = h^\lambda_{11} = 0$, that is, $A_\lambda = 0$ for all λ. On the other hand, we see that h^1_{11} is a constant. Thus it will be easily verified that the second fundamental forms of M are parallel. Therefore we have the following

Lemma 11.1. Let M be an anti-invariant surface of a Sasakian space form $\bar{M}^{2m+1}(c)$. If the mean curvature vector is parallel, then M is flat and M is an anti-invariant submanifold in an $\bar{M}^3(c)$ in $\bar{M}^{2m+1}(c)$, where $\bar{M}^3(c)$ is a totally geodesic invariant submanifold of $\bar{M}^{2m+1}(c)$.

If we take S^{2m+1} as the ambient manifold, then the normal connection of M is flat. Thus M is an anti-invariant surface of an S^3 in S^{2m+1}. Therefore, by Theorem 6.4, we have

Theorem 11.1. Let M be a compact orientable anti-invariant surface with parallel mean curvature vector of S^{2m+1}. Then

$$M = S^1(r_1) \times S^1(r_2) \qquad \text{in an } S^3 \text{ in } S^{2m+1}$$

where $r_1^2 + r_2^2 = 1$.

Theorem 11.2. Let M be a compact orientable anti-invariant minimal surface of S^{2m+1}. Then

$$M = S^1(\frac{1}{\sqrt{2}}) \times S^1(\frac{1}{\sqrt{2}}) \qquad \text{in an } S^3 \text{ in } S^{2m+1}.$$

From Examples 8.1, 8.2 and Lemma 11.1 we have

Theorem 11.3. Let M be a complete anti-invariant surface with parallel mean curvature vector of $E^{2m+1}(-3)$. Then M is E^2 in an $E^3(-3)$ in $E^{2m+1}(-3)$ or M is a product of the form

$$M = S^1(r_1) \times E^1 \qquad \text{in an } E^3(-3) \text{ in } E^{2m+1}(-3).$$

Theorem 11.4. Let M be a complete anti-invariant minimal surface of $E^{2m+1}(-3)$. Then M is E^2 in an $E^3(-3)$ in $E^{2m+1}(-3)$.

§12. Anti-invariant submanifolds of a Sasakian manifold with vanishing contact Bochner curvature tensor

Let $\bar{M}$ be a (2m+1)-dimensional Sasakian manifold. As an analogue of the Bochner curvature tensor in a Kaehlerian manifold, we define the contact Bochner curvature tensor in a Sasakian manifold $\bar{M}$ by

(Matsumoto-Chūman [39] and Yano [71])

$$(12.1)\qquad B_{ABCD} = K_{ABCD} + (\delta_{AC} - \eta_A\eta_C)L_{BD} - (\delta_{AD} - \eta_A\eta_D)L_{BC}$$

$$+ L_{AC}(\delta_{BD} - \eta_B\eta_D) - L_{AD}(\delta_{BC} - \eta_B\eta_C)$$

$$+ \phi_{AC}M_{BD} - \phi_{AD}M_{BC} + M_{AC}\phi_{BD} - M_{AD}\phi_{BC}$$

$$+ 2(M_{AB}\phi_{CD} + \phi_{AB}M_{CD}) + (\phi_{AC}\phi_{BD} - \phi_{AD}\phi_{BC} + 2\phi_{AB}\phi_{CD}),$$

where we have put

$$(12.2)\qquad L_{BD} = \frac{1}{2(m+2)}[- K_{BD} - (L+3)\delta_{BD} + (L-1)\eta_B\eta_D],$$

$$(12.3)\qquad L = \sum_A L_{AA} = - \frac{K+2(3m+2)}{4(m+1)},$$

$$(12.4)\qquad M_{BD} = - \sum_C L_{BC}\phi_{DC},$$

K_{BD} and K denoting the Ricci tensor and the scalar curvature of $\bar{M}$ respectively. We easily see that $L_{BD} = L_{DB}$ and $M_{BD} = - M_{DB}$.

By a straightforward computation we can prove (cf. Matsumoto-Chūman [39] and Yano [71])

$$(12.5)\qquad \sum_A \bar{\nabla}_A B_{ABCD} = - 2m[\bar{\nabla}_C L_{BD} - \bar{\nabla}_D L_{CB} - \eta_C(\phi_{BD} + M_{BD})$$

$$+ \eta_D(\phi_{BC} + M_{BC}) - 2\eta_B(\phi_{CD} + M_{CD})$$

$$- \frac{1}{2(m+2)} \sum_A (\phi_{AC}\phi_{BD} - \phi_{AD}\phi_{BC} + 2\phi_{AB}\phi_{CD})(\bar{\nabla}_A K)].$$

Let M be an $(n+1)$-dimensional anti-invariant submanifold of $\bar{M}$. Then, from (12.1), we find

$$B_{ijkl} = K_{ijkl} + (\delta_{ik} - \eta_i\eta_k)L_{jl} - (\delta_{il} - \eta_i\eta_l)L_{jk} + L_{ik}(\delta_{jl} - \eta_j\eta_l) - L_{il}(\delta_{jk} - \eta_j\eta_k). \tag{12.6}$$

Now, we rewrite the Gauss equation (3.7) of Chapter II in the form

$$K_{ijkl} = R_{ijkl} - D_{ijkl}, \tag{12.7}$$

where we have put

$$D_{ijkl} = \sum_a (h^a_{ik}h^a_{jl} - h^a_{il}h^a_{jk}). \tag{12.8}$$

Then (12.6) becomes

$$B_{ijkl} = R_{ijkl} - D_{ijkl} + (\delta_{ik} - \eta_i\eta_k)L_{jl} - (\delta_{il} - \eta_i\eta_l)L_{jk} + L_{ik}(\delta_{jl} - \eta_j\eta_l) - L_{il}(\delta_{jk} - \eta_j\eta_k). \tag{12.9}$$

Here we notice that the structure vector field ξ is parallel and M is locally a Riemannian direct product of M^n and M^1 generated by ξ and M^n is totally geodesic in M. Therefore, in M^n, we have

$$B_{xyzv} = R_{xyzv} - D_{xyzv} + \delta_{xz}L_{yv} - \delta_{xv}L_{yz} + L_{xz}\delta_{yv} - L_{xv}\delta_{yz}, \tag{12.10}$$

where R_{xyzv} is the Riemannian curvature tensor of M^n.

Now we put

$$D_{xy} = \sum_z D_{xzyz}, \qquad D = \sum_x D_{xx}, \qquad b_{xy} = \sum_z B_{xzyz},$$

$$b = \sum_x b_{xx}, \qquad G = \sum_x L_{xx}.$$

From (12.10), we obtain by contraction

$$(12.11) \qquad b_{yv} = R_{yv} - D_{yv} + (n-2)L_{yv} + G\delta_{yv},$$

$$(12.12) \qquad b = r - D + 2(n-1)G,$$

$R_{yv} = \sum_x R_{xyxv}$ and $r = \sum_x R_{xx}$ being the Ricci tensor and the scalar curvature of M^n respectively. From (12.11) and (12.12) we have

$$(12.13) \qquad L_{yv} = -\frac{1}{(n-2)}(R_{yv} - D_{yv} - b_{yv}) + \frac{1}{2(n-1)(n-2)}(r - D - b)\delta_{yv}.$$

Substituting (12.13) into (12.10) we find

$$\begin{aligned}(12.14) \qquad B_{xyzv} = C_{xyzv} - D_{xyzv} &+ \frac{1}{(n-2)}(\delta_{xz}b_{yv} + \delta_{yv}b_{xz} - \delta_{xv}b_{yz} \\ &- \delta_{yz}b_{xv} + \delta_{xz}D_{yv} + \delta_{yv}D_{xz} - \delta_{xv}D_{yz} - \delta_{yz}D_{xv}) \\ &- \frac{1}{(n-1)(n-2)}(\delta_{xz}\delta_{yv} - \delta_{xv}\delta_{yz})(b + D),\end{aligned}$$

where C_{xyzv} denotes the Weyl conformal curvature tensor of M^n.

Lemma 12.1. Let M be an (n+1)-dimensional ($n \geq 4$) anti-invariant submanifold of a (2m+1)-dimensional Sasakian manifold $\bar{M}$ with vanishing contact Bochner curvature tensor. If

$$(12.15) \qquad D_{xyzv} = \alpha(\delta_{xz}\delta_{yv} - \delta_{xv}\delta_{yz})$$

for some scalar function α on M, then M is locally a Riemannian direct product $M^n \times M^1$, where M^n is a totally geodesic, conformally flat hypersurface in M and M^1 is a 1-dimensional space generated by ξ.

Proof. Since $\bar{M}$ has vanishing contact Bochner curvature tensor, we have

$$B_{xyzv} = 0, \qquad b_{yv} = 0, \qquad b = 0.$$

Thus (12.14) becomes

$$C_{xyzv} = D_{xyzv} - \frac{1}{(n-2)}(\delta_{xz}D_{yv} + \delta_{yv}D_{xz} - \delta_{xv}D_{yz} - \delta_{yz}D_{xv}) + \frac{1}{(n-1)(n-2)}(\delta_{xz}\delta_{yv} - \delta_{xv}\delta_{yz})D. \tag{12.16}$$

On the other hand, from the assumption (12.15) we find, by contraction,

$$D_{yv} = \alpha(n-1)\delta_{yv}, \qquad D = \alpha n(n-1). \tag{12.17}$$

Substituting (12.17) into (12.16) we find $C_{xyzv} = 0$ and consequently, n being greater than 3, M^n is conformally flat.

Theorem 12.1 (Yano [72]). Let M be an (n+1)-dimensional ($n \geq 4$) anti-invariant submanifold of a (2m+1)-dimensional Sasakian manifold $\bar{M}$ with vanishing contact Bochner curvature tensor. If the second fundamental forms of M are commutative and if the f-structure in the normal bundle is parallel, then M is locally a product of a conformally flat Riemannian manifold M^n and a 1-dimensional space M^1.

Proof. Since the second fundamental forms are commutative, we have

$$\begin{aligned} D_{xyzv} &= \sum_w (h^w_{xz}h^w_{yv} - h^w_{xv}h^w_{yz}) = \sum_w (h^x_{wz}h^y_{wv} - h^x_{wv}h^y_{wz}) \\ &= -(\delta_{xz}\delta_{yv} - \delta_{xv}\delta_{yz}). \end{aligned} \tag{12.18}$$

From this and Lemma 12.1 we see that our assertion holds.

Theorem 12.2. Let M be an (n+1)-dimensional ($n \geq 4$) anti-invariant submanifold of a (2m+1)-dimensional Sasakian manifold $\bar{M}$ with vanishing contact Bochner curvature tensor. If the f-structure in the normal bundle is parallel and if $H_aH_b = H_bH_a$ for all a and b, then M is locally a product of a conformally flat Riemannian manifold M^n and a 1-dimensional space M^1.

Proof. From the assumption we easily see that $D_{xyzv} = 0$. Thus, by Lemma 12.1, M^n is conformally flat.

Theorem 12.3 (Kon [31]). Let M be an (n+1)-dimensional ($n \geq 4$) H-totally umbilical, anti-invariant submanifold of a (2m+1)-dimensional Sasakian manifold $\bar{M}$ with vanishing contact Bochner curvature tensor. Then M is locally a product of a conformally flat Riemannian manifold M^n and a 1-dimensional space M^1.

Proof. From the assumption we have

$$D_{xyzv} = \sum_a (\mathrm{Tr}H_a)^2/n^2(\delta_{xz}\delta_{yv} - \delta_{xv}\delta_{yz}). \tag{12.19}$$

From (12.19) and Lemma 12.1 we have our assertion.

In the sequel, we study the case that dim M = 4, i.e., n = 3.

Theorem 12.4 (Kon [31]). Let M be a 4-dimensional anti-invariant submanifold of a (2m+1)-dimensional Sasakian manifold $\bar{M}$ with vanishing contact Bochner curvature tensor. If M is H-totally umbilical, then M is locally a product of a conformally flat Riemannian manifold M^3 and a 1-dimensional space M^1.

Proof. If we put $U_a = (\mathrm{Tr}H_a)/n$, then (12.13) becomes

$$L_{yv} = C_{yv} + \frac{1}{2}\sum_a U_a^2\delta_{yv}, \tag{12.20}$$

where

$$C_{yv} = -\frac{1}{(n-2)}R_{yv} + \frac{1}{2(n-1)(n-2)}r\delta_{yv}.$$

On the other hand, the Codazzi equation (2.5) of Chapter II implies

(12.21) $$\sum_x K_{axyx} = \sum_x (h^a_{xxy} - h^a_{xyx}) = (n-1)\nabla_y U_a.$$

By (12.1) we also have

(12.22) $$\sum_x K_{axyx} = -(n-1)L_{ay} + 3\sum_{C,x}\phi_{ax}L_{yC}\phi_{xC}.$$

From (12.21) and (12.22) we find

(12.23) $$\nabla_y U_a = -L_{ay} + \frac{3}{(n-1)}\sum_{C,x}\phi_{ax}L_{yC}\phi_{xC}.$$

From the assumption and Corollary 8.2 we see that $\sum_a \phi_{ax}U_a = U_{x*} = 0$. Thus we have

(12.24) $$\sum_a (\nabla_y U_a)U_a = -\sum_a L_{ay}U_a.$$

From (12.20) we have

(12.25) $$\bar{\nabla}_x L_{yv} + \sum_a h^a_{xy}L_{av} + \sum_a h^a_{xv}L_{ya} = \nabla_x C_{yv} + \sum_a (\nabla_x U_a)U_a \delta_{yv}.$$

Using (12.24) and (12.25), we find

(12.26) $$\bar{\nabla}_x L_{yv} - \bar{\nabla}_y L_{xv} = \nabla_x C_{yv} - \nabla_y C_{xv}.$$

By the assumption and (12.5) we have $\nabla_x C_{yv} = \nabla_y C_{xv}$, which shows that M^3 is conformally flat.

Chapter V

Anti-invariant submanifolds of Sasakian manifolds normal to the structure vector field

From Proposition 7.3 of Chapter II, we see that if a submanifold of a Sasakian manifold is normal to the structure vector field, then the submanifold is anti-invariant. So that we mean, in this chapter, by an anti-invariant submanifold M of a Sasakian manifold $\bar{M}$ a submanifold M of $\bar{M}$ normal to the structure vector field ξ of $\bar{M}$.

§1. Fundamental properties

Let $\bar{M}$ be a (2m+1)-dimensional Sasakian manifold with structure tensors $(\phi,\xi,\eta,\bar{g})$ and let M be an n-dimensional anti-invariant submanifold of $\bar{M}$. We choose a local field of orthonormal frames $e_1,\ldots,e_n;e_{n+1},\ldots,e_m;e_{0^*}=\xi$, $e_{1^*}=\phi e_1,\ldots,e_{n^*}=\phi e_n;e_{(n+1)^*}=\phi e_{n+1},\ldots,e_{m^*}=\phi e_m$ in $\bar{M}$ in such a way that, restricted to M, $e_1,\ldots,e_n$ are tangent to M. With respect to this frame field of $\bar{M}$, let $\omega^1,\ldots,\omega^n;\omega^{n+1},\ldots,\omega^m;\omega^{0^*},\omega^{1^*},\ldots,\omega^{n^*};\omega^{(n+1)^*},\ldots,\omega^{m^*}$ be the dual forms. Unless otherwise stated, we use the conventions that the ranges of indices are respectively:

$$A, B, C, D = 1,\ldots,m,0^*,1^*,\ldots,m^*,$$

$$i, j, k, l, t, s = 1,\ldots,n,$$

$$a, b, c, d = n+1,\ldots,m,0^*,1^*,\ldots,m^*,$$

$$x, y, z = 0^*,1^*,\ldots,n^*,$$

$$\alpha, \beta, \gamma = n+1,\ldots,m,$$

$$p, q, r = n+1,\ldots,m,1^*,\ldots,m^*,$$

$$\lambda, \mu, \nu = n+1,\ldots,m,0^*,(n+1)^*,\ldots,m^*.$$

Then we have the following equations:

$$(1.1)\qquad \begin{aligned} &\omega_j^i = \omega_{j^*}^{i^*}, \quad \omega_j^{i^*} = \omega_i^{j^*}, \quad \omega^i = \omega_{0^*}^{i^*}, \quad \omega^{i^*} = -\omega_{0^*}^{i}, \\ &\omega_\beta^\alpha = \omega_{\beta^*}^{\alpha^*}, \quad \omega_\beta^{\alpha^*} = \omega_\alpha^{\beta^*}, \quad \omega^\alpha = \omega_{0^*}^{\alpha^*}, \quad \omega^{\alpha^*} = -\omega_{0^*}^{\alpha}, \\ &\omega_\alpha^i = \omega_{\alpha^*}^{i^*}, \quad \omega_\alpha^{i^*} = \omega_i^{\alpha^*}. \end{aligned}$$

Restricting these forms to M, we have $\omega^a = 0$ and hence $\omega^{i^*} = \omega_i^{0^*} = h_{ij}^0\omega^j = 0$. From this and (1.1) we find

$$(1.2)\qquad h_{jk}^i = h_{ik}^j = h_{ij}^k, \qquad h_{ij}^0 = 0,$$

where we have used h_{jk}^i in place of $h_{jk}^{i^*}$ to simplify the notation.

Lemma 1.1. Let M be an n-dimensional anti-invariant submanifold of a (2m+1)-dimensional Sasakian manifold $\bar{M}$. If the second fundamental forms of M are parallel, then we have

$$(1.3)\qquad A_t = 0, \qquad \text{i.e.,} \qquad h_{ij}^t = 0.$$

Proof. By the assumption, (3.10) of Chapter II and (1.2), we have $h_{ijk}^0 = -h_{ij}^t = 0$.

Since $\phi T_x(M) \subset T_x(M)^\perp$ at each point x of M, we have the decomposition of $T_x(M)^\perp$ into the direct sum

$$T_x(M)^\perp = \phi T_x(M) \oplus N_x(M),$$

where $N_x(M)$ is the orthogonal complement of $\phi T_x(M)$ in the normal space $T_x(M)^\perp$. If a vector $N \in N_x(M)$, then we have $\phi N \in N_x(M)$. We also have $\xi \in N_x(M)$.

For any vector field N in the normal bundle $T(M)^{\perp}$, we put

$$(1.4) \qquad \phi N = PN + fN,$$

where PN is the tangential part of ϕN and fN the normal part of ϕN. Then P is a tangent bundle valued 1-form on the normal bundle and f is an endomorphism of the normal bundle. Putting $N = \phi X$ in (1.4) and applying ϕ to (1.4), we find

$$(1.5) \qquad \begin{aligned} PfN &= 0, \qquad & f^2N &= -N - \phi PN + \eta(N)\xi, \\ P\phi X &= -X, \qquad & f\phi X &= 0 \end{aligned}$$

for any tangent vector field X to M and for any normal vector field N to M. From (1.5) we find

$$f^3 + f = 0.$$

If f does not vanish, f being of constant rank, it defines an f-structure in the normal bundle. Using the Gauss and Weingarten formulas, we have, from (1.4),

$$(1.6) \qquad (D_X f)N = -B(X,PN) - \phi A_N X.$$

If $D_X f = 0$ for all X, then the f-structure in the normal bundle is said to be parallel.

Lemma 1.2. Let M be an n-dimensional anti-invariant submanifold of a (2m+1)-dimensional Sasakian manifold $\bar{M}$. If the f-structure in the normal bundle is parallel, then we have

$$(1.7) \qquad A_N = 0 \quad \text{for} \quad N \in N_x(M), \qquad \text{i.e.,} \quad h_{ij}^{\lambda} = 0.$$

Proof. From the assumption and (1.6) we have $\phi A_N X = 0$. Consequently we have

$$\phi^2 A_N X = - A_N X + \eta(A_N X)\xi = - A_N X = 0,$$

which proves (1.7).

From Lemmas 1.1 and 1.2 we have

Lemma 1.3. Let M be an n-dimensional anti-invariant submanifold of a (2m+1)-dimensional Sasakian manifold $\bar{M}$. If the f-structure in the normal bundle is parallel and if the second fundamental forms of M are parallel, then M is totally geodesic.

From (3.10) of Chapter II and (1.2) we obtain

$$\sum_i h^0_{iit} = - \sum_i h^t_{ii}. \tag{1.8}$$

If the mean curvature vector of M is parallel, then (1.8) implies that $\sum_i h^t_{ii} = 0$ for all t. From this and Lemma 1.2 we have

Lemma 1.4. Let M be an n-dimensional anti-invariant submanifold of a (2m+1)-dimensional Sasakian manifold $\bar{M}$. If the f-structure in the normal bundle is parallel and if the mean curvature vector of M is parallel, then M is minimal.

Lemma 1.5. Let M be an n-dimensional anti-invariant submanifold of a (2n+1)-dimensional Sasakian manifold $\bar{M}$. If the mean curvature vector of M is parallel, then M is minimal.

From Lemmas 1.4 and 1.5 we see that the notion of parallel mean curvature vector is not essential for anti-invariant submanifolds. Thus we need the following notion. On an anti-invariant submanifold M of a Sasakian manifold $\bar{M}$, if $\sum_i h^p_{iit} = 0$ for all p and t, then we say that the mean curvature vector of M is η-parallel.

On the other hand, in view of Lemma 1.1 we need the following notion. The second fundamental forms of M are said to be η-parallel if $h^p_{ijk} = 0$ for all p, i, j and k.

In the sequel, we assume that the ambient manifold $\bar{M}$ is of constant φ-sectional curvature c and denote it by $\bar{M}^{2m+1}(c)$. The curvature tensor K of $\bar{M}^{2m+1}(c)$ is given by (2.3) of Chapter IV. Let M be an n-dimensional anti-invariant submanifold of $\bar{M}^{2m+1}(c)$. Then the Gauss equation (3.7) of Chapter II and (2.3) of Chapter IV imply

$$(1.9) \qquad R^i_{jkl} = \frac{1}{4}(c+3)(\delta_{ik}\delta_{jl} - \delta_{il}\delta_{jk}) + \sum_a (h^a_{ik}h^a_{jl} - h^a_{il}h^a_{jk}),$$

from which

$$(1.10) \qquad R_{ij} = \frac{1}{4}(n-1)(c+3)\delta_{ij} + \sum_{a,k} (h^a_{ii}h^a_{jj} - h^a_{ij}h^a_{ij})$$

and

$$(1.11) \qquad r = \frac{1}{4}n(n-1)(c+3) + \sum_{a,i,j} (h^a_{ii}h^a_{jj} - h^a_{ij}h^a_{ij}).$$

From (1.9) we have

Proposition 1.1. Let M be an n-dimensional anti-invariant submanifold of a Sasakian space form $\bar{M}^{2m+1}(c)$. If M is totally geodesic, then M is of constant curvature $\frac{1}{4}(c+3)$.

If M is minimal, by (1.10) and (1.11), we have respectively

$$(1.12) \qquad R_{ij} = \frac{1}{4}(n-1)(c+3)\delta_{ij} - \sum_{a,k} h^a_{ik}h^a_{jk},$$

$$(1.13) \qquad r = \frac{1}{4}n(n-1)(c+3) - \sum_{a,i,j} (h^a_{ij})^2.$$

Proposition 1.2. Let M be an n-dimensional anti-invariant minimal submanifold of a Sasakian space form $\bar{M}^{2m+1}(c)$. Then

(a) $S - \frac{1}{4}(n-1)(c+3)g$ is negative semi-definite,

(b) $r \leq \frac{1}{4}n(n-1)(c+3)$,

S being the Ricci tensor and r the scalar curvature of M.

Proposition 1.3. Let M be an n-dimensional anti-invariant minimal submanifold of a Sasakian space form $\bar{M}^{2m+1}(c)$. Then M is totally geodesic if and only if M satisfies one of the following conditions:

(a) M is of constant curvature $\frac{1}{4}(c+3)$,

(b) $S = \frac{1}{4}(n-1)(c+3)g$,

(c) $r = \frac{1}{4}n(n-1)(c+3)$.

If the f-structure in the normal bundle is parallel, then (1.9), (1.10) and (1.11) can be respectively rewritten as

$$R^i_{jkl} = \frac{1}{4}(c+3)(\delta_{ik}\delta_{jl} - \delta_{il}\delta_{jk}) + \sum_t (h^t_{ik}h^t_{jl} - h^t_{il}h^t_{jk}), \tag{1.14}$$

$$R_{ij} = \frac{1}{4}(n-1)(c+3)\delta_{ij} + \sum_{t,k}(h^t_{kk}h^t_{ij} - h^t_{ik}h^t_{jk}) \tag{1.15}$$

and

$$r = \frac{1}{4}n(n-1)(c+3) + \sum_{t,i,j}(h^t_{ii}h^t_{jj} - h^t_{ij}h^t_{ij}). \tag{1.16}$$

If M is moreover minimal, then we have

$$R_{ij} = \frac{1}{4}(n-1)(c+3)\delta_{ij} - \sum_{t,k} h^t_{ik}h^t_{jk}, \tag{1.17}$$

$$r = \frac{1}{4}n(n-1)(c+3) - \sum_{t,i,j}(h^t_{ij})^2. \tag{1.18}$$

§2. Connection in the normal bundle

Let M be an n-dimensional anti-invariant submanifold of a (2m+1)-dimensional Sasakian manifold $\bar{M}$. Then the curvature tensor K of $\bar{M}$ satisfies

$$(2.1) \qquad K^{i}_{jkl} = K^{i^*}_{j^*kl} + (\delta_{ik}\delta_{jl} - \delta_{il}\delta_{jk}).$$

If the f-structure in the normal bundle is parallel, then the Gauss equation (3.7) of Chapter II and (1.3) give

$$(2.2) \qquad R^{i}_{jkl} = K^{i}_{jkl} + \sum_{t}(h^{t}_{ik}h^{t}_{jl} - h^{t}_{il}h^{t}_{jk}).$$

On the other hand, using (3.9) of Chapter II, we have

$$(2.3) \qquad R^{i^*}_{j^*kl} = K^{i^*}_{j^*kl} + \sum_{t}(h^{i}_{tk}h^{j}_{tl} - h^{i}_{tl}h^{j}_{tk}).$$

From (1.2), (2.2) and (2.3) we obtain

$$(2.4) \qquad R^{i^*}_{j^*kl} - K^{i^*}_{j^*kl} = R^{i}_{jkl} - K^{i}_{jkl}.$$

Substituting (2.1) into (2.4), we find

$$(2.5) \qquad R^{i^*}_{j^*kl} = R^{i}_{jkl} - (\delta_{ik}\delta_{jl} - \delta_{il}\delta_{jk}),$$

from which we have

Proposition 2.1. Let M be an n-dimensional anti-invariant submanifold of a (2m+1)-dimensional Sasakian manifold $\bar{M}$. If the f-structure in the normal bundle is parallel and the normal connection of M is flat, then M is of constant curvature 1.

Corollary 2.1. Let M be an n-dimensional anti-invariant submanifold of a (2n+1)-dimensional Sasakian manifold $\bar{M}$. Then the normal connection of M is flat if and only if M is of constant curvature 1.

§3. Computation of Laplacian

Let M be an n-dimensional anti-invarint submanifold of a Sasakian space form $\bar{M}^{2m+1}(c)$. Since the second fundamental forms of M satisfy the Codazzi equation $h^a_{ijk} - h^a_{ikj} = 0$, using (3.18) of Chapter II and (2.3) of Chapter IV, we find

$$(3.1)\qquad \sum_{a,i,j} h^a_{ij}\Delta h^a_{ij} = \sum_{a,i,j,k} h^a_{ij}h^a_{kkij} + \frac{1}{4}(c+3)\sum_a [n\mathrm{Tr}A_a^2 - (\mathrm{Tr}A_a)^2]$$
$$+ \frac{1}{4}(c-1)\sum_t [\mathrm{Tr}A_t^2 - (\mathrm{Tr}A_t)^2] + \sum_{a,b} \{\mathrm{Tr}(A_aA_b - A_bA_a)^2$$
$$- [\mathrm{Tr}(A_aA_b)]^2 + \mathrm{Tr}A_b\mathrm{Tr}(A_aA_bA_a)\}.$$

We now put

$$S_{ab} = \sum_{i,j} h^a_{ij}h^b_{ij}, \qquad S_a = S_{aa}, \qquad S = \sum_a S_a.$$

From (1.2) we see that $S_{a0} = 0$ for all a. On the other hand, from (3.10) of Chapter II and (1.2), we have

$$(3.2)\qquad \sum_{a,i,j} h^a_{ij}\Delta h^a_{ij} = \frac{1}{2}\Delta S - \sum_{a,i,j,k} (h^a_{ijk})^2$$
$$= \frac{1}{2}\Delta S - \sum_{p,i,j,k} (h^p_{ijk})^2 - \sum_t \mathrm{Tr}A_t^2.$$

From (3.1) and (3.2) we obtain

Lemma 3.1. Let M be an n-dimensional anti-invariant submanifold of a Sasakian space form $\bar{M}^{2m+1}(c)$. Then we have

$$(3.3)\qquad \frac{1}{2}\Delta S - \sum_{p,i,j,k}(h^p_{ijk})^2 = \sum_{a,i,j,k} h^a_{ij}h^a_{kkij} + \frac{1}{4}(c+3)nS - \sum_a S_a^2$$

$$- \frac{1}{4}(c+3)\sum_a (TrA_a)^2 + \frac{1}{4}(c+3)\sum_t S_t - \frac{1}{4}(c-1)\sum_t (TrA_t)^2$$

$$+ \sum_{a,b} [Tr(A_aA_b - A_bA_a)^2 + TrA_bTr(A_aA_bA_a)].$$

Lemma 3.2. Let M be an n-dimensional anti-invariant submanifold of a Sasakian space form $\bar{M}^{2m+1}(c)$. If the f-structure in the normal bundle is parallel, then we have

$$(3.4)\qquad \frac{1}{2}\Delta S - \sum_{p,i,j,k}(h^p_{ijk})^2 = \sum_{a,i,j,k} h^a_{ij}h^a_{kkij} + \frac{1}{4}(n+1)(c+3)S - \sum_t S_t^2$$

$$- \frac{1}{2}(c+1)\sum_t (TrA_t)^2 + \sum_{t,s} [Tr(A_tA_s - A_sA_t)^2 + TrA_sTr(A_tA_sA_t)].$$

Lemma 3.3. Let M be an n-dimensional anti-invariant submanifold of a Sasakian space form $\bar{M}^{2m+1}(c)$. If M is minimal, then we have

$$(3.5)\qquad \frac{1}{2}\Delta S - \sum_{p,i,j,k}(h^p_{ijk})^2 = \frac{1}{4}n(c+3)S + \frac{1}{4}(c+3)\sum_t S_t - \sum_a S_a^2$$

$$+ \sum_{a,b} Tr(A_aA_b - A_bA_a)^2.$$

Lemma 3.4. Let M be an n-dimensional anti-invariant submanifold with parallel mean curvature vector of a Sasakian space form $\bar{M}^{2m+1}(c)$. If the f-structure in the normal bundle is parallel, then we have

$$(3.6)\qquad \frac{1}{2}\Delta S - \sum_{p,i,j,k}(h^p_{ijk})^2 = \frac{1}{4}(n+1)(c+3)S - \sum_t S_t^2$$

$$+ \sum_{t,s} Tr(A_tA_s - A_sA_t)^2.$$

Proof. From Lemma 1.4 we see that M is minimal. Therefore equation

(3.5) gives (3.6)

Lemma 3.5. Let M be an n-dimensional anti-invariant submanifold with parallel mean curvature vector of a Sasakian space form $\bar{M}^{2m+1}(c)$. If the f-structure in the normal bundle is parallel, then we have

$$(3.7)\qquad \sum_{p,i,j,k} (h^p_{ijk})^2 - \frac{1}{2}\Delta S \leq [(2 - \frac{1}{n})S - \frac{1}{4}(n+1)(c+3)]S.$$

Proof. Applying Lemma 3.5 of Chapter III, we obtain

$$(3.8)\qquad - \sum_{t,s} Tr(A_tA_s - A_sA_t)^2 + \sum_t S_t^2 - \frac{1}{4}(n+1)(c+3)S$$

$$\leq 2 \sum_{t\neq s} S_tS_s + \sum_t S_t^2 - \frac{1}{4}(n+1)(c+3)S$$

$$= [(2 - \frac{1}{n})S - \frac{1}{4}(n+1)(c+3)]S - \frac{1}{n}\sum_{t>s} (S_t - S_s)^2.$$

From (3.6) and (3.8) we have (3.7).

If M is compact and orientable, inequality (3.7) implies the following

Theorem 3.1. Let M be an n-dimensional compact orientable anti-invariant submanifold with parallel mean curvature vector of a Sasakian space form $\bar{M}^{2m+1}(c)$. If the f-structure in the normal bundle is parallel, then we have the integral inequality

$$(3.9)\qquad \int_M [(2 - \frac{1}{n})S - \frac{1}{4}(n+1)(c+3)]S*1 \geq \int_M \sum_{p,i,j,k} (h^p_{ijk})^2*1 \geq 0.$$

Corollary 3.1. Let M be an n-dimensional compact orientable anti-invariant submanifold with parallel mean curvature vector of a Sasakian space form $\bar{M}^{2m+1}(c)$. If the f-structure in the normal bundle is parallel, then either M is totally geodesic, or $S = n(n+1)(c+3)/4(2n-1)$ or at some point x of M,

$S(x) > n(n+1)(c+3)/4(2n-1)$.

Proof. Suppose $S \leq n(n+1)(c+3)/4(2n-1)$ everywhere on M. Then (3.9) implies that the second fundamental forms of M are η-parallel and hence S is a constant. Therefore, either $S = 0$, that is, M is totally geodesic, or $S = n(n+1)(c+3)/4(2n-1)$. Except these posibilities, $S(x) > n(n+1)(c+3)/4(2n-1)$ at some point x of M.

§4. Anti-invariant surfaces of S^5

In this section we study anti-invariant submanifolds satisfying $S = n(n+1)(c+3)/4(2n-1)$. In the following we take as the ambient manifold $\bar{M}$, that of constant curvature 1, that is, we assume that $c = 1$.

Theorem 4.1. Let M be an n-dimensional ($n > 1$) anti-invariant submanifold with parallel mean curvature vector of a Sasakian space form $\bar{M}^{2m+1}(1)$. If the f-structure in the normal bundle is parallel and $S = n(n+1)/(2n-1)$, then $n = 2$ and M is a flat anti-invariant minimal surface of some $\bar{M}^5(1)$ in $\bar{M}^{2m+1}(1)$, where $\bar{M}^5(1)$ is a totally geodesic invariant submanifold of dimension 5 of $\bar{M}^{2m+1}(1)$. With respect to an adapted dual orthonormal frame field $\omega^1, \omega^2, \omega^{0*}, \omega^{1*}, \omega^{2*}$ in $\bar{M}^5(1)$, the connection forms (ω^A_B) of $\bar{M}^5(1)$, restricted to M, are given by

$$\begin{pmatrix} 0 & 0 & 0 & -\lambda\omega^2 & -\lambda\omega^1 \\ 0 & 0 & 0 & -\lambda\omega^1 & \lambda\omega^2 \\ 0 & 0 & 0 & -\omega^1 & -\omega^2 \\ \lambda\omega^2 & \lambda\omega^1 & \omega^1 & 0 & 0 \\ \lambda\omega^1 & -\lambda\omega^2 & \omega^2 & 0 & 0 \end{pmatrix}, \qquad \lambda = \frac{1}{\sqrt{2}} .$$

Proof. First of all, by the assumption, M is a minimal submanifold. Moreover, from the assumption on the square of the length of the second fundamental form of M, we see that S is a constant. Therefore (3.7) implies that the second fundamental form of M is η-parallel. By Lemma 3.5 of Chapter III and (3.8) we have

$$\sum_{t>s} (S_t - S_s)^2 = 0, \tag{4.1}$$

$$- \mathrm{Tr}(A_t A_s - A_s A_t)^2 = 2\mathrm{Tr}A_t^2 \mathrm{Tr}A_s^2. \tag{4.2}$$

Therefore we have $S_t = S_s$ and we can assume that $A_t = 0$ for $t = 3,\ldots,n$. Thus we must have $n = 2$. Then we obtain

$$A_1 = \lambda \begin{pmatrix} 0 & 1 \\ 1 & 0 \end{pmatrix}, \qquad A_2 = \lambda \begin{pmatrix} 1 & 0 \\ 0 & -1 \end{pmatrix}, \qquad A_\lambda = 0, \tag{4.3}$$

by using $h_{12}^1 = \lambda = h_{11}^2$ and (1.7). Thus we have

$$\omega_i^\lambda = 0. \tag{4.4}$$

Since the second fundamental form of M is η-parallel, using (3.10) of Chapter II, we have

$$dh_{ij}^p = h_{ik}^p \omega_j^k + h_{kj}^p \omega_i^k - h_{ij}^q \omega_q^p. \tag{4.5}$$

Since $h_{ij}^\lambda = 0$, we can rewrite (4.5) as

$$dh_{ij}^t = h_{ik}^t \omega_j^k + h_{kj}^t \omega_i^k - h_{ij}^s \omega_{s*}^{t*}. \tag{4.6}$$

From (4.5) we find $h^s_{ij}\omega^\lambda_{s*} = 0$ and hence we have

$$\omega^\lambda_{t*} = 0. \tag{4.7}$$

Putting t = 1, i = 1 and j = 2 in (4.6), we see that $d\lambda = 0$, which means that λ is a constant. Since S = 2, we have $2\lambda^2 = 1$. Thus we may assume that $\lambda = 1/\sqrt{2}$. Moreover, since M is not totally geodesic, i.e., $\lambda \neq 0$, we have

$$\omega_i^{t*} \neq 0. \tag{4.8}$$

Since we have (4.4), (4.7) and (4.8) we define a distribution L by

$$\omega^\lambda = 0, \qquad \omega_i^\lambda = 0, \qquad \omega_x^\lambda = 0.$$

It easily follows from the structure equations that

$$d\omega^\lambda = 0, \qquad d\omega_i^\lambda = 0, \qquad d\omega_x^\lambda = 0.$$

Therefore L is a 5-dimensional completely integrable distribution. We consider the maximal integral submanifold $\bar{M}(x)$ of L throgth $x \in M$. Then $\bar{M}(x)$ is of dimension 5 and by construction it is a totally geodesic invariant submanifold of $\bar{M}^{2m+1}(1)$ and hence it is of constant curvature 1. Moreover M is an anti-invariant submanifold of $\bar{M}^5(1)$. In $\bar{M}^5(1)$, (4.3) and (4.6) imply the equations:

$$\omega_1^{1*} = \lambda\omega^2, \qquad \omega_2^{1*} = \omega_1^{2*} = \lambda\omega^1, \qquad \omega_2^{2*} = -\lambda\omega^2,$$

$$\omega_{0*}^{1*} = \omega^1, \qquad \omega_{0*}^{2*} = \omega^2, \qquad \omega_{0*}^1 = \omega_{0*}^2 = 0, \tag{4.9}$$

$$\omega_1^2 = \omega_{1*}^{2*} = 0.$$

From the Gauss equation (3.7) of Chapter II and (4.3) we easily see that M is flat. These considerations prove our assertion.

Example 4.1. Let J be the almost complex structure of C^{n+1} given by

$$J = \begin{pmatrix} \begin{array}{cc|} 0 & -1 \\ 1 & 0 \\ \hline \end{array} & & & \\ & \ddots & & \\ & & \ddots & \\ & & & \begin{array}{|cc} \hline 0 & -1 \\ 1 & 0 \end{array} \end{pmatrix} .$$

Let S^{2n+1} be a (2n+1)-dimensional unit sphere in C^{n+1} with standard Sasakian structure $(\phi,\xi,\eta,\bar{g})$. Let S^1 be a circle of radius 1. Let us consider

$$T^n = S^1 \times \ldots\ldots \times S^1.$$

Then we can construct an isometric minimal immersion of T^n into S^{2n+1} which is anti-invariant in the following way.

Let $X: T^n \longrightarrow S^{2n+1}$ be a minimal immersion represented by

$$X = \frac{1}{n+1}(\cos u^1, \sin u^1, \ldots\ldots, \cos u^n, \sin u^n, \cos u^{n+1}, \sin u^{n+1}),$$

where we have put $u^{n+1} = -(u^1 + \ldots + u^n)$. We may regard X as a position vector of S^{2n+1} in C^{n+1}. The structure vector field ξ of S^{2n+1}, restricted to T^n, is then given by

$$\xi = JX = \frac{1}{n+1}(-\sin u^1, \cos u^1, \ldots, -\sin u^{n+1}, \cos u^{n+1}).$$

Putting $X_i = \partial X/\partial u^i$, we have

$$X_i = \frac{1}{n+1}(0,\ldots,0,-\sin u^i,\cos u^i,0,\ldots,0,\sin u^{n+1},-\cos u^{n+1}),$$

where $i = 1,\ldots,n$. Thus X_i, $i = 1,\ldots,n$, are linearly independent and $\eta(X_i) = 0$ for $i = 1,\ldots,n$. Therefore the immersion X is anti-invariant.

Let $M^{n+1} = S^1(r) \times \ldots\ldots \times S^1(r)$, where $r = 1/\sqrt{n+1}$. Then M^{n+1} is an anti-invariant submanifold of S^{2n+1} tangent to the structure vector field (see Example 6.2 of Chapter IV). By the definition of the immersion X, T^n is an anti-invariant flat minimal submanifold of S^{2n+1} by the immersion X and the square of the length of the second fundamental form of T^n is equal to $S = n(n-1)$.

Especially a torus $T = S^1 \times S^1$ is an anti-invariant surface of S^5 by the immersion X. For this case we have $S = 2$. Obviously, T^n is an anti-invariant submanifold of S^{2m+1} $(m > n)$ with parallel f-structure in the normal bundle.

From Theorem 4.1 and Example 4.1 we have

Theorem 4.2. Let M be an n-dimensional $(n > 1)$ compact orientable anti-invariant submanifold with parallel mean curvature vector and with parallel f-structure in the normal bundle of S^{2m+1}. If $S = n(n+1)/(2n-1)$, then M is $S^1 \times S^1$ in an S^5 in S^{2m+1}.

We now prove a pinching theorem with respect to the square of the length of the second fundamental form.

Theorem 4.3. Let M be an n-dimensional compact orientable anti-invariant minimal submanifold of a Sasakian space form $\bar{M}^{2m+1}(c)$ $(c > -3)$. If $S \leq \frac{1}{4}(c+3)nq/(2q-1)$, then M is totally geodesic, where $q = 2m-n$.

Proof. From Lemma 3.5 of Chapter III and (3.5) we have

$$\sum_{p,i,j,k} (h^p_{ijk})^2 - \frac{1}{2}\Delta S \leq [(2 - \frac{1}{n})S - \frac{1}{4}(c+3)n]S - \frac{1}{4}(c+3)\sum_t S_t - \frac{1}{q}\sum_{p>r} (S_p - S_r)^2. \tag{4.10}$$

From the assumption and (4.10) we find

(4.11) $$\sum_{p>r}(S_p - S_r)^2 = 0,$$

(4.12) $$\sum_t S_t = 0.$$

Therefore we have $S_p = S_r$ for all p and r and $S_t = 0$ for all t. Thus we must have $S_p = 0$, that is, $A_p = 0$ for all p. Since $A_0 = 0$, we obtain $A_a = 0$ for all a. Consequently M is totally geodesic.

§5. η-parallel mean curvature vector

We first prove the following lemmas.

Lemma 5.1. Let M be an n-dimensional anti-invariant submanifold of a (2m+1)-dimensional Sasakian manifold $\bar{M}$. If the second fundamental forms of M are commutative, then we can choose an orthonormal frame for which A_t is of the form

$$A_t = \begin{pmatrix} 0 & & & & & & \\ & \ddots & & & & & \\ & & 0 & & & & \\ & & & \lambda_t & & & \\ & & & & 0 & & \\ & & & & & \ddots & \\ & & & & & & 0 \end{pmatrix} t, \qquad t = 1,\dots,n,$$

that is, $h^t_{ij} = 0$ unless $t = i = j$. Moreover, if the f-structure in the normal bundle is parallel, then the second fundamental forms of M are commutative if and only if we can choose an orthonormal frame for which $h^t_{ij} = 0$ unless $t = i = j$.

Proof. Since $A_aA_b = A_bA_a$, we can choose an orthonormal frame $e_1,\dots,e_n$ for which all A_a are simultaneously diagonal, i.e., $h^a_{ij} = 0$ when $i \neq j$, that is, $h^t_{ij} = 0$ when $i \neq j$. From (1.2) we see that $h^t_{ij} = 0$

unless $t = i = j$. When the f-structure in the normal bundle is parallel, Lemma 1.2 proves our assertion.

Corollary 5.1. Let M be an n-dimensional anti-invariant submanifold with parallel mean curvature vector of a (2m+1)-dimensional Sasakian manifold $\bar{M}$. If the f-structure in the normal bundle is parallel and if the second fundamental forms of M are commutative, then M is totally geodesic.

Proof. From Lemma 1.5 we see that M is minimal. Then we have $\mathrm{Tr}A_t = 0$ for all t. Thus Lemma 5.1 implies $\lambda_t = 0$ and hence $A_t = 0$. Thus taking account of Lemma 1.2, we have our assertion.

Corollary 5.2. Let M be an n-dimensional ($n > 1$) anti-invariant submanifold with parallel f-structure in the normal bundle of a (2m+1)-dimensional Sasakian manifold $\bar{M}$. If M is totally umbilical, then M is totally geodesic.

Proof. From the assumption we see that the second fundamental forms of M are commutative. Thus $h^t_{ij} = 0$ unless $t = i = j$ by Lemma 5.1. On the other hand, we get $h^t_{ij} = \lambda_t \delta_{ij}/n$. Putting $i = j \neq t$, we have $\lambda_t = 0$, that is, $A_t = 0$. Since $A_\lambda = 0$, M is totally geodesic.

From Proposition 1.1, Corollaries 5.1 and 5.2 we have

Corollary 5.3. Let M be an n-dimensional anti-invariant submanifold with parallel mean curvature vector of a Sasakian space form $\bar{M}^{2m+1}(c)$. If the f-structure in the normal bundle is parallel and the second fundamental forms of M are commutative, then M is of constant curvature $\frac{1}{4}(c+3)$.

Corollary 5.4. Let M be an n-dimensional ($n > 1$) anti-invariant submanifold with parallel f-structure in the normal bundle of a Sasakian space form $\bar{M}^{2m+1}(c)$. If M is totally umbilical, then M is of constant curvature $\frac{1}{4}(c+3)$.

Lemma 5.2. Let M be an n-dimensional anti-invariant submanifold with parallel f-structure in the normal bundle of a Sasakian space form $\bar{M}^{2m+1}(c)$. Then M is of constant curvature $\frac{1}{4}(c+3)$ if and only if the second fundamental forms of M are commutative.

Proof. From (1.2) and (1.4) we have

$$(5.1) \qquad R^i_{jkl} = \frac{1}{4}(c+3)(\delta_{ik}\delta_{jl} - \delta_{il}\delta_{jk}) + \sum_t (h^i_{tk}h^j_{tl} - h^i_{tl}h^j_{tk}),$$

which proves our assertion.

From (1.2) we have the following lemma.

Lemma 5.3. Let M be an n-dimensional anti-invariant submanifold of a (2m+1)-dimensional Sasakian manifold $\bar{M}$. Then we have

$$(5.2) \qquad \sum_{t,s} \mathrm{Tr}A_t^2A_s^2 = \sum_{t,s} (\mathrm{Tr}A_tA_s)^2.$$

Lemma 5.4. Let M be an n-dimensional anti-invariant submanifold of constant curvature k of a Sasakian space form $\bar{M}^{2m+1}(c)$. If the f-structure in the normal bundle is parallel, then we have

$$(5.3) \qquad [\frac{1}{4}(c+3)-k]\sum_t [\mathrm{Tr}A_t^2 - (\mathrm{Tr}A_t)^2] = \sum_{t,s} [\mathrm{Tr}A_t^2A_s^2 - \mathrm{Tr}(A_tA_s)^2].$$

Proof. From (1.14) we have

$$(5.4) \qquad [\frac{1}{4}(c+3)-k](\delta_{ik}\delta_{jl} - \delta_{il}\delta_{jk}) = \sum_t (h^t_{il}h^t_{jk} - h^t_{ik}h^t_{jl}).$$

Multiplying both sides of (5.4) by $\sum_s h^s_{il}h^s_{jk}$, summing up with respect to i, j, k and l and using (5.2), we have (5.3).

Lemma 5.5. Let M be an n-dimensional anti-invariant submanifold of constant curvature k of a Sasakian space form $\bar{M}^{2m+1}(c)$. If the f-structure in the normal bundle is parallel, then we have

$$(5.5)\qquad (n-1)[\tfrac{1}{4}(c+3)-k]\sum_t TrA_t^2 = \sum_{t,s}[TrA_t^2A_s^2 - TrA_s Tr(A_tA_sA_t)].$$

Proof. From (5.4) we obtain

$$(5.6)\qquad (n-1)[\tfrac{1}{4}(c+3)-k]\delta_{jl} = \sum_{t,i}(h_{il}^t h_{ij}^t - h_{ii}^t h_{jl}^t).$$

Multiplying both sides of (5.6) by $\sum_s h_{jk}^s h_{kl}^s$ and summing up with respect to j, k and l, we have (5.5).

Lemma 5.6. Let M be an n-dimensional anti-invariant submanifold with η-parallel mean curvature vector of a Sasakian space form $\bar{M}^{2m+1}(c)$. Then the square of the length of the second fundamental form of M is constant, i.e., S = constant if and only if the scalar curvature r of M is constant.

Proof. From the assumption and (1.11) we have our assertion.

Lemma 5.7. Let M be an n-dimensional anti-invariant submanifold with η-parallel mean curvature vector and with parallel f-structure in the normal bundle of a Sasakian space form $\bar{M}^{2m+1}(c)$. If M is of constant curvature k, then we have

$$(5.7)\qquad \sum_{p,i,j,k}(h_{ijk}^p)^2 = -k\sum_t[(n+1)TrA_t^2 - 2(TrA_t)^2].$$

Proof. First of all, from (3.10) of Chapter II and (1.2), we have

$$-\sum_{t,i,j,k} h_{ij}^t h_{kkij}^t = \sum_t (TrA_t)^2.$$

Substituting this into (3.4) and using Lemma 5.6, we obtain

$$(5.8)\qquad \sum_{p,i,j,k}(h^p_{ijk})^2 = -\sum_t[\tfrac{1}{4}(n+1)(c+3)\mathrm{Tr}A_t^2 - \tfrac{1}{2}(c+3)(\mathrm{Tr}A_t)^2]$$

$$-\sum_{t,s}\{\mathrm{Tr}(A_tA_s - A_sA_t)^2 - [\mathrm{Tr}(A_tA_s)]^2 + \mathrm{Tr}A_s\mathrm{Tr}(A_tA_sA_t)\}.$$

Substituting (5.3) and (5.5) into (5.8) and using (5.2) we have (5.7).

Proposition 5.1. Let M be an n-dimensional anti-invariant submanifold with η-parallel mean curvature vector of a Sasakian space form $\bar{M}^{2m+1}(c)$. If the f-structure in the normal bundle is parallel and M is flat, then the second fundamental form of M is η-parallel.

Proof. If M is flat, (5.7) implies $h^p_{ijk} = 0$, which means that the second fundamental form of M is η-parallel.

Theorem 5.1. Let M be an n-dimensional $(n > 1)$ anti-invariant submanifold with η-parallel mean curvature vector and with parallel f-structure in the normal bundle of a Sasakian space form $\bar{M}^{2m+1}(c)$. If M is of constant curvature k and if $\frac{1}{4}(c+3) \geq k$, then either $k \leq 0$ or M is totally geodesic and $\frac{1}{4}(c+3) = k$.

Proof. From (5.6) we obtain

$$[\tfrac{1}{4}(c+3)-k]n(n-1) = \sum_t[\mathrm{Tr}A_t^2 - (\mathrm{Tr}A_t)^2].$$

By the assumption we have $\sum_t \mathrm{Tr}A_t^2 \geq \sum_t(\mathrm{Tr}A_t)^2$. Thus if $k > 0$, then (5.7) implies

$$0 = \sum_t\{(n-1)\mathrm{Tr}A_t^2 + 2[\mathrm{Tr}A_t^2 - (\mathrm{Tr}A_t)^2]\}.$$

Since $n > 1$, we have $\sum_t \mathrm{Tr}A_t^2 = 0$ and hence M is totally geodesic. Except this possibility, we have $k \leq 0$.

Theorem 5.2. Let M be an n-dimensional ($n > 1$) anti-invariant submanifold of constant curvature k with η-parallel second fundamental form of a Sasakian space form $\bar{M}^{2m+1}(c)$. If the f-structure in the normal bundle is parallel and $\frac{1}{4}(c+3) \geq k$, then either M is totally geodesic or flat.

Lemma 5.8. Let M be an n-dimensional anti-invariant submanifold with parallel mean curvature vector and of constant curvature k of a Sasakian space form $\bar{M}^{2m+1}(c)$. If the f-structure in the normal bundle is parallel, then we have

$$\sum_{p,i,j,k} (h^p_{ijk})^2 = -k(n+1)S. \tag{5.9}$$

Proof. By the assumption M is minimal. Therefore (5.7) gives (5.9).

Corollary 5.5. Let M be an n-dimensional anti-invariant submanifold with parallel mean curvature vector and of constant curvature k of a Sasakian space form $\bar{M}^{2m+1}(c)$. If the f-structure in the normal bundle is parallel, then either $k \leq 0$ or M is totally geodesic.

Corollary 5.6. Let M be an n-dimensional anti-invariant submanifold with parallel mean curvature vector and of constant curvature k of a Sasakian space form $\bar{M}^{2m+1}(c)$. If the f-structure in the normal bundle is parallel and the second fundamental form of M is η-parallel, then either M is totally geodesic or flat.

Lemma 5.9. Let M be an n-dimensional anti-invariant submanifold with η-parallel mean curvature vector of a Sasakian space form $\bar{M}^{2m+1}(c)$. If the f-structure in the normal bundle is parallel and the second fundamental forms of M are commutative, then we have

$$\sum_{p,i,j,k} (h^p_{ijk})^2 = -\frac{1}{4}(c+3)(n-1)S. \tag{5.10}$$

Proof. Using Lemmas 5.1 and 5.2, we can transform (5.7) into (5.10).

Theorem 5.3. Let M be an n-dimensional $(n > 1)$ anti-invariant submanifold with η-parallel mean curvature vector and with commutative second fundamental forms of a Sasakian space form $\bar{M}^{2m+1}(c)$. If the f-structure in the normal bundle is parallel, then either M is totally geodesic or $c \leq -3$.

Theorem 5.4. Let M be an n-dimensional $(n > 1)$ anti-invariant submanifold with η-parallel second fundamental forms and with commutative second fundamental forms of a Sasakian space form $\bar{M}^{2m+1}(c)$. If the f-structure in the normal bundle is parallel, then either M is totally geodesic or flat.

Proof. By the assumption and Lemma 5.2, M is of constant curvature $\frac{1}{4}(c+3)$. On the other hand, by (5.10), M is either totally geodesic or $c = -3$, in which case M is flat.

§6. Anti-invariant submanifolds of $E^{2m+1}(-3)$

First of all, we state some examples of anti-invariant submanifolds of a Sasakian space form $E^{2m+1}(-3)$.

Example 6.1. Let $E^{2m+1}(-3)$ be a Sasakian space form of constant ϕ-sectional curvature -3 as in Example 2 of Chapter I. We denote by E^n an n-dimensional Euclidean space. Then we can construct a natural imbedding E^n into $E^{2n+1}(-3)$ by

$$(x^1,\ldots,x^n) \longmapsto (x^1,\ldots,x^n,0,\ldots,0,0),$$

where $(x^1,\ldots,x^n)$ is a coordinate system of E^n. Then E^n is an anti-invariant submanifold of $E^{2n+1}(-3)$ and hence of $E^{2m+1}(-3)$ $(m > n)$. Obviously E^n is totally geodesic in $E^{2m+1}(-3)$.

Example 6.2. Let $S^1(r_i) = \{z_i \in C : |z_i|^2 = r_i^2\}$ be a circle of radius r_i, $i = 1,\ldots,n$. We consider the following pythagorean product

$$M^n = S^1(r_1) \times \cdots\cdots \times S^1(r_n).$$

We construct an imbedding of M^n into $E^{2n+1}(-3)$ by

$$(r_1\cos u^1,\ldots,r_n\cos u^n,r_1\sin u^1,\ldots,r_n\sin u^n,0).$$

Then M^n is an n-dimensional anti-invariant submanifold of $E^{2n+1}(-3)$ and hence of $E^{2m+1}(-3)$ $(m > n)$. On the other hand, M^n is a totally geodesic submanifold of $S^1(r_1) \times \cdots\cdots \times S^1(r_n) \times E$ as in Example 8.2 of Chapter IV. Moreover the mean curvature vector of M^n is η-parallel and the second fundamental forms of M^n in $E^{2n+1}(-3)$ are given by

$$A_t = \begin{pmatrix} 0 & & & & & & \\ & \ddots & & & & & \\ & & 0 & & & & \\ & & & \lambda_t & & & \\ & & & & 0 & & \\ & & & & & \ddots & \\ & & & & & & 0 \end{pmatrix} \begin{matrix} \\ \\ \\ t, \\ \\ \\ \\ \end{matrix} \qquad t = 1,\ldots,n.$$

Similarly we see that

$$M^n = S^1(r_1) \times \cdots\cdots \times S^1(r_p) \times E^{n-p}, \qquad 1 \leq p \leq n,$$

is a complete anti-invariant submanifold of $E^{2n+1}(-3)$ with η-parallel mean curvature vector whose second fundamental forms are given by

$$A_t = \begin{pmatrix} 0 & & & & & & \\ & \ddots & & & & & \\ & & 0 & & & & \\ & & & \lambda_t & & & \\ & & & & 0 & & \\ & & & & & \ddots & \\ & & & & & & 0 \end{pmatrix} \begin{matrix} \\ \\ \\ t, \\ \\ \\ \\ \end{matrix} \qquad t = 1,\ldots,p, \qquad A_s = 0, \qquad s = p+1,\ldots,n.$$

From Lemma 5.2, Proposition 5.1 and Theorem 5.4, combined with Example 6.1 and Example 6.2, we have

Theorem 6.1. Let M be an n-dimensional complete anti-invariant submanifold with η-parallel mean curvature vector and with parallel f-structure in the normal bundle of $E^{2m+1}(-3)$. If the second fundamental forms of M are commutative, then M is totally geodesic, that is, E^n in an $E^{2n+1}(-3)$ in $E^{2m+1}(-3)$, or a pythagorean product of the form

$$S^1(r_1) \times \cdots \times S^1(r_p) \times E^{n-p} \quad \text{in an } E^{2n+1}(-3) \text{ in } E^{2m+1}(-3),$$

where $1 \leq p < n$, or a product of the form

$$S^1(r_1) \times \cdots \times S^1(r_n) \quad \text{in an } E^{2n+1}(-3) \text{ in } E^{2m+1}(-3).$$

From Theorem 5.4 and Theorem 6.1 we obtain

Theorem 6.2. Let M be an n-dimensional ($n > 1$) complete anti-invariant submanifold with η-parallel second fundamental form and with parallel f-structure in the normal bundle of a simply connected complete Sasakian space form $\bar{M}^{2m+1}(c)$. If the second fundamental forms of M are commutative, then M is totally geodesic, or a product of the form

$$S^1(r_1) \times \cdots \times S^1(r_p) \times E^{n-p} \quad \text{in an } E^{2n+1}(-3) \text{ in } E^{2m+1}(-3),$$

where $1 \leq p < n$, or a product of the form

$$S^1(r_1) \times \cdots \times S^1(r_n) \quad \text{in an } E^{2n+1}(-3) \text{ in } E^{2m+1}(-3).$$

Theorem 6.3. Under the same assumptions as in Theorem 6.1, if M is compact, then M is a product of the form

$$S^1(r_1) \times \cdots \times S^1(r_n) \quad \text{in an } E^{2n+1}(-3) \text{ in } E^{2m+1}(-3).$$

Theorem 6.4. Under the same assumptions as in Theorem 6.2, if M is compact, then M is a product of the form

$$S^1(r_1) \times \ldots\ldots \times S^1(r_n) \quad \text{in an } E^{2n+1}(-3) \text{ in } E^{2m+1}(-3).$$

§7. η-parallel second fundamental form

Let M be an n-dimensional anti-invariant submanifold of a Sasakian space form $\bar{M}^{2m+1}(c)$. We assume that the f-structure in the normal bundle is parallel and the mean curvature vector of M is parallel. Then M is a minimal submanifold. From (1.2), (1.14), (1.17) and (1.18) we have

$$(7.1) \qquad \sum_{t,s}(R_{ts})^2 = \frac{1}{n}r^2 - \frac{1}{n}S^2 + \sum_{t,s}\mathrm{Tr}A_t^2A_s^2,$$

$$(7.2) \qquad \sum_{i,j,k,l}(R^i_{jkl})^2 = \frac{2}{n(n-1)}r^2 - \frac{2}{n(n-1)}S^2 - \sum_{t,s}\mathrm{Tr}(A_tA_s - A_sA_t)^2.$$

Now we assume that the second fundamental form of M is η-parallel. Then (3.6) implies

$$(7.3) \qquad \frac{1}{4}(n+1)(c+3)S + \sum_{t,s}[\mathrm{Tr}(A_tA_s - A_sA_t)^2 - \mathrm{Tr}A_t^2A_s^2] = 0.$$

From this we obtain

Proposition 7.1. Let M be an n-dimensional anti-invariant submanifold with η-parallel second fundamental form and with parallel mean curvature vector of a Sasakian space form $\bar{M}^{2m+1}(c)$. If the f-structure in the normal bundle is parallel and $c \leq -3$, then M is totally geodesic.

Substituting (7.1) and (7.2) into (7.3) we have

$$(7.4) \qquad \frac{(n+1)}{n(n-1)}rS = \sum_{i,j,k,l}(R^i_{jkl})^2 - \frac{2}{n(n-1)}r^2 + \sum_{k,l}(R_{kl})^2 - \frac{1}{n}r^2.$$

We see that the right hand side of (7.4) is non-negative and if it vanishes, then M is of constant curvature. Thus, if the scalar curvature r of M vanishes, then M is flat. Consequently we have

Theorem 7.1. Let M be an n-dimensional anti-invariant submanifold with parallel mean curvature vector and with parallel f-structure in the normal bundle of a Sasakian space form $\bar{M}^{2m+1}(c)$. If the second fundamental form of M is η-parallel, then either M is totally geodesic or M has the non-negative scalar curvature $r \geq 0$. Moreover if $r = 0$, then M is flat.

Corollary 7.1. Let M be an n-dimensional anti-invariant submanifold with parallel mean curvature vector of a Sasakian space form $\bar{M}^{2n+1}(c)$. If the second fundamental form of M is η-parallel, then either M is totally geodesic or M has the non-negative scalar curvature $r \geq 0$. Moreover if $r = 0$, then M is flat.

§8. Minimal surfaces

Let M be an anti-invariant surface with parallel mean curvature vector of a Sasakian space form $\bar{M}^5(c)$. Then M is a minimal surface. If we make a suitable choice of a frame, the second fundamental forms of M are given by

$$(8.1) \qquad A_0 = 0, \quad A_1 = \begin{pmatrix} a & 0 \\ 0 & -a \end{pmatrix}, \quad A_2 = \begin{pmatrix} 0 & -a \\ -a & 0 \end{pmatrix}.$$

From (3.6) and (8.1) we find

$$(8.2) \qquad \frac{1}{2}\Delta S = \sum_{t,i,j,k} (h^t_{ijk})^2 + 3(c+3)a^2 - 24a^4.$$

On the other hand, the Gauss curvature K of M is given by

$$K = \frac{1}{4}(c+3) - 2a^2. \tag{8.3}$$

From (8.2) and (8.3) we have

$$\frac{1}{2}\Delta S = \sum_{t,i,j,k} (h_{ijk}^t)^2 + 12a^2K, \tag{8.4}$$

from which

$$\Delta \log S = 6K. \tag{8.5}$$

From this, we obtain, using the method of section 9 of Chapter III,

Theorem 8.1 (Yamaguchi-Kon-Miyahara [63]). Let M be an anti-invariant surface with parallel mean curvature vector of a Sasakian space form $\bar{M}^5(c)$.

(1) If M has genus zero, then M is totally geodesic.

(2) If M is a complete surface with non-negative curvature, then M is totally geodesic or flat.

(3) If M is a complete surface with non-positive Gauss curvature K and if $\frac{1}{4}(c+3)-K \geq \alpha > 0$ for some constant α, then M is flat.

§9. Anti-invariant submanifolds normal to ξ of a Sasakian manifold with vanishing contact Bochner curvature tensor

Let $\bar{M}$ be a (2m+1)-dimensional Sasakian manifold and M be an n-dimensional anti-invariant submanifold of $\bar{M}$. Then, by (12.1) of Chapter IV, we have

$$B_{ijkl} = K_{ijkl} + \delta_{ik}L_{jl} - \delta_{il}L_{jk} + L_{ik}\delta_{jl} - L_{il}\delta_{jk}. \tag{9.1}$$

If we put $D_{ijkl} = \sum_a (h_{ik}^a h_{jl}^a - h_{il}^a h_{jk}^a)$, (9.1) becomes

$$(9.2)\qquad B_{ijkl} = R_{ijkl} - D_{ijkl} + \delta_{ik}L_{jl} - \delta_{il}L_{jk} + L_{ik}\delta_{jl} - L_{il}\delta_{jk}.$$

Hereafter we put

$$D_{ij} = \sum_k D_{ikjk},\qquad D = \sum_i D_{ii},\qquad b_{ij} = \sum_k B_{ikjk},$$

$$b = \sum_i b_{ii},\qquad G = \sum_i L_{ii}.$$

By contraction, we find, from (9.2),

$$(9.3)\qquad b_{ij} = R_{ij} - D_{ij} + (n-2)L_{ij} + G\delta_{ij},$$

$$(9.4)\qquad b = r - D + 2(n-1)G.$$

From (9.3) and (9.4) we see that

$$(9.5)\qquad L_{ij} = -\frac{1}{(n-2)}(R_{ij} - D_{ij} - b_{ij}) + \frac{1}{2(n-1)(n-2)}(r - D - b)\delta_{ij}.$$

Substituting (9.5) into (9.2) we obtain

$$(9.6)\qquad \begin{aligned} B_{ijkl} = C_{ijkl} - D_{ijkl} &+ \frac{1}{(n-2)}(\delta_{ik}b_{jl} + \delta_{jl}b_{ik} - \delta_{il}b_{jk} - \delta_{jk}b_{il} \\ &+ \delta_{ik}D_{jl} + \delta_{jl}D_{ik} - \delta_{il}D_{jk} - \delta_{jk}D_{il}) \\ &- \frac{1}{(n-1)(n-2)}(\delta_{ik}\delta_{jl} - \delta_{il}\delta_{jk})(b + D), \end{aligned}$$

where C_{ijkl} denotes the Weyl conformal curvature tensor of M.

Lemma 9.1. Let M be an n-dimensional ($n \geq 4$) anti-invariant submanifold of a (2m+1)-dimensional Sasakian manifold $\bar{M}$ with vanishing contact Bochner curvature tensor. If

$$(9.7)\qquad D_{ijkl} = \alpha(\delta_{ik}\delta_{jl} - \delta_{il}\delta_{jk})$$

for some scalar function α, then M is conformally flat.

Proof. Since $\bar{M}$ has vanishing contact Bochner curvature tensor, we have

$$B_{ijkl} = 0, \qquad b_{ij} = 0, \qquad b = 0.$$

Therefore (9.6) becomes

$$(9.8) \qquad C_{ijkl} = D_{ijkl} - \frac{1}{(n-2)}(\delta_{ik}D_{jl} + \delta_{jl}D_{ik} - \delta_{il}D_{jk} - \delta_{jk}D_{il}) + \frac{1}{(n-1)(n-2)}(\delta_{ik}\delta_{jl} - \delta_{il}\delta_{jk})D.$$

On the other hand, from the assumption (9.7), we find, by contraction,

$$(9.9) \qquad D_{ij} = \alpha(n-1)\delta_{ij}, \qquad D = \alpha n(n-1).$$

Substituting (9.9) into (9.8), we have $C_{ijkl} = 0$ and consequently M is conformally flat.

Theorem 9.1 (Yano [72]). Let M be an n-dimensional ($n \geq 4$) anti-invariant submanifold of a (2m+1)-dimensional Sasakian manifold $\bar{M}$ with vanishing contact Bochner curvature tensor. If the second fundamental forms of M are commutative and if the f-structure in the normal bundle is parallel, then M is conformally flat.

Proof. From the assumption we have

$$D_{ijkl} = \sum_t (h^t_{ik}h^t_{jl} - h^t_{il}h^t_{jk}) = \sum_t (h^i_{tk}h^j_{tl} - h^i_{tl}h^j_{tk}) = 0.$$

From this and Lemma 9.1, we see that M is conformally flat.

Theorem 9.2 (Yano [72]). Let M be an n-dimensional ($n \geq 3$) totally umbilical, anti-invariant submanifold of a (2m+1)-dimensional Sasakian manifold $\bar{M}$ with vanishing contact Bochner curvature tensor. Then M is conformally flat.

Proof. From the assumption we have

$$D_{ijkl} = [\sum_a (\mathrm{Tr}A_a)^2/n^2](\delta_{ik}\delta_{jl} - \delta_{il}\delta_{jk}).$$

From this and Lemma 9.1, we see that if $n \geq 4$, then M is conformally flat.

When $n = 3$, by the similar method as in Theorem 12.4 of Chapter IV, using the Codazzi equation (2.5) of Chapter II and the fact $A_t = 0$ (see the proof of Corollary 5.2), we find

$$\bar{\nabla}_i L_{jk} - \bar{\nabla}_j L_{ik} = \nabla_i C_{jk} - \nabla_j C_{ik}.$$

Then, by virtue of (12.5) of Chapter IV, the left hand side of this equation vanishes. Therefore M is conformally flat.

Chapter VI

Anti-invariant submanifolds and Riemannian fibre bundles

In this chapter we study relations between submanifolds of Kaehlerian manifolds and those of Sasakian manifolds. For this purpose we use the method of Riemannian fibre bundles (cf. Ishihara-Konishi [17], Lawson [35], O'Neill [49] and Yano-Ishihara [74]). As an application we study certain real hypersurfaces of a complex projective space CP^m.

§1. Regular Sasakian manifolds

Let $\bar{M}$ be a (2m+1)-dimensional Sasakian manifold. If $\bar{M}$ is regular, then there is a fibering $\bar{\pi} : \bar{M} \longrightarrow \bar{M}/\xi = \bar{N}$, where $\bar{N}$ denotes the set of orbits of ξ and is a real 2m-dimensional Kaehlerian manifold (cf. Sasaki [50]). Let $(\phi,\xi,\eta,\bar{g})$ be the Sasakian structure of $\bar{M}$ and $(J,\bar{G})$ the Kaehlerian structure of $\bar{N}$. We denote by * the horizontal lift with respect to the connection η. Then we have

$$(1.1)\qquad (JX)^* = \phi X^*, \qquad \bar{g}(X^*,Y^*) = \bar{G}(X,Y)$$

for any vector fields X and Y on $\bar{N}$. We denote by $\bar{\nabla}$ (resp. $\bar{\nabla}'$) the operator of covariant differentiation with respect to $\bar{g}$ (resp. $\bar{G}$). Then we obtain

$$(1.2)\qquad (\bar{\nabla}'_X Y)^* = -\phi^2\bar{\nabla}_{X^*}Y^*.$$

Since $\bar{M}$ is a Sasakian manifold, equation (1.2) can be rewritten as

$$(1.3)\qquad (\bar{\nabla}'_X Y)^* = \bar{\nabla}_{X^*}Y^* + \bar{g}(Y^*,\phi X^*)\xi.$$

We denote by $\bar{R}$ and $\bar{R}'$ the Riemannian curvature tensors of $\bar{M}$ and $\bar{N}$ respectively. Then we have

Lemma 1.1. The Riemannian curvature tensors $\bar{R}$ and $\bar{R}'$ satisfy

$$(1.4)\qquad (\bar{R}'(X,Y)Z)^* = \bar{R}(X^*,Y^*)Z^* + \bar{g}(Z^*,\phi Y^*)\phi X^*$$
$$- \bar{g}(Z^*,\phi X^*)\phi Y^* - 2\bar{g}(Y^*,\phi X^*)\phi Z^*$$

for any vector fields X, Y and Z on $\bar{N}$.

Proof. From (1.3) we have

$$(\bar{\nabla}'_X\bar{\nabla}'_Y Z)^* = \bar{\nabla}_{X^*}(\bar{\nabla}'_Y Z)^* + \bar{g}((\bar{\nabla}'_Y Z)^*,\phi X^*)\xi$$
$$= \bar{\nabla}_{X^*}\bar{\nabla}_{Y^*}Z^* + \bar{g}(\bar{\nabla}_{X^*}Z^*,\phi Y^*)\xi + \bar{g}(\bar{\nabla}_{Y^*}Z^*,\phi X^*)\xi$$
$$+ \bar{g}(Z^*,\phi\bar{\nabla}_{X^*}Y^*)\xi + \bar{g}(Z^*,\phi Y^*)\phi X^*.$$

Similarly, we get

$$(\bar{\nabla}'_Y\bar{\nabla}'_X Z)^* = \bar{\nabla}_{Y^*}\bar{\nabla}_{X^*}Z^* + \bar{g}(\bar{\nabla}_{Y^*}Z^*,\phi X^*)\xi + \bar{g}(\bar{\nabla}_{X^*}Z^*,\phi X^*)\xi$$
$$+ \bar{g}(Z^*,\phi\bar{\nabla}_{Y^*}X^*)\xi + \bar{g}(Z^*,\phi X^*)\phi Y^*.$$

Moreover, we have

$$(\bar{\nabla}_{[X,Y]}Z)^* = \bar{\nabla}_{[X^*,Y^*]}Z^* + \bar{g}(Z^*,\phi[X^*,Y^*])\xi + 2\bar{g}(Y^*,\phi X^*)\phi Z^*.$$

From these equations we have (1.4).

Let $\bar{S}$ and $\bar{S}'$ be the Ricci tensors of $\bar{M}$ and $\bar{N}$ respectively. Then, from (1.4) we have

$$(1.5)\qquad \bar{S}'(X,Y) = \bar{S}(X^*,Y^*) + 2\bar{g}(X^*,Y^*),$$

from which

(1.6) $$\bar{r}' = \bar{r} + 2m,$$

where $\bar{r}$ and $\bar{r}'$ are the scalar curvatures of $\bar{M}$ and $\bar{N}$ respectively. Moreover, the sectional curvatures of $\bar{M}$ and $\bar{N}$ determined by orthonormal vectors X and Y on $\bar{N}$ satisfy

(1.7) $$\bar{K}'(X,Y) = \bar{K}(X^*,Y^*) + 3\bar{g}(X^*,\phi Y^*)^2,$$

from which

(1.8) $$\bar{K}'(X,JY) = \bar{K}(X^*,\phi X^*) + 3.$$

Theorem 1.1. $\bar{M}$ is of constant ϕ-sectional curvature c if and only if $\bar{N}$ is of constant holomorphic sectional curvature (c+3).

Proof. Since the structure vector field ξ is a vertical vector field, (7.2) of Chapter I and (1.4) imply

(1.9) $$(4\bar{R}'(X,Y)Z)^* = (c+3)[\bar{g}(Y^*,Z^*)X^* - \bar{g}(X^*,Z^*)Y^* + \bar{g}(Z^*,\phi Y^*)\phi X^* - \bar{g}(Z^*,\phi X^*)\phi Y^* - 2\bar{g}(Y^*,\phi X^*)\phi Z^*].$$

On the other hand, from (1.1) and (1.9) we have

(1.10) $$4\bar{R}'(X,Y)Z = (c+3)[\bar{G}(Y,Z)X - \bar{G}(X,Z)Y + \bar{G}(Z,JY)JX - \bar{G}(Z,JX)JY - 2\bar{G}(Y,JX)JZ].$$

Thus, if $\bar{M}$ is of constant ϕ-sectional curvature c, then $\bar{N}$ is of constant holomorphic sectional curvature (c+3). Conversely, if $\bar{N}$ is of constant holomorphic sectional curvature (c+3), then (1.4) and (1.9) imply that $\bar{M}$ is of constant ϕ-sectional curvature c.

In the next place, we study the relation between contact Bochner curvature tensor $\bar{B}$ of $\bar{M}$ and Bochner curvature tensor $\bar{B}'$ of $\bar{N}$. We put

$$(1.11) \qquad \bar{L}(X,Y) = \frac{1}{2(m+2)}[-\bar{S}(X,Y) - (\bar{L}+3)\bar{g}(X,Y) + (\bar{L}-1)\eta(X)\eta(Y)],$$

$$(1.12) \qquad \bar{L} = \sum_{i=1}^{2m+1} \bar{L}(e_i,e_i) = -\frac{\bar{r}+2(3m+2)}{4(m+1)},$$

$$(1.13) \qquad \bar{M}(X,Y) = -\bar{L}(X,\phi Y)$$

for any vectors X, Y on $\bar{M}$, where $\{e_i\}$, $i = 1,\dots,2m+1$, denotes an orthonormal frame of $\bar{M}$ (see (12.2), (12.3) and (12.4) in §12 of Chapter IV).

The corresponding objects of $\bar{N}$ are given by

$$(1.14) \qquad \bar{L}(X,Y) = \frac{1}{2(m+2)}[-\bar{S}'(X,Y) + \frac{1}{(m+1)}\bar{r}'\bar{G}(X,Y)],$$

$$(1.15) \qquad \bar{L}' = \sum_{i=1}^{2m} \bar{L}'(e_i,e_i) = -\frac{\bar{r}'}{4(m+1)},$$

$$(1.16) \qquad \bar{M}'(X,Y) = -\bar{L}'(X,JY)$$

for any vectors X, Y on $\bar{N}$, where $\{e_i\}$, $i = 1,\dots,2m$, denotes an orthonormal frame of $\bar{N}$ (see (10.2) and (10.3) in §10 of Chapter III).

Using (1.5) and (1.6), we obtain

$$(1.17) \qquad (\bar{L}'(X,Y))^* = \bar{L}(X^*,Y^*),$$

$$(1.18) \qquad (\bar{M}'(X,Y))^* = \bar{M}(X^*,Y^*).$$

From (1.17), (1.18), (10.1) of Chapter III and (12.1) of Chapter IV, we have

$$(1.19) \qquad \bar{G}(\bar{B}'(X,Y)Z,W)^* = \bar{g}(\bar{B}(X^*,Y^*)Z^*,W^*)$$

for any vector fields X, Y, Z and W on $\bar{N}$. On the other hand, we easily see that

(1.20) $$\bar{B}(X,\xi)Y = 0$$

for any vector fields X and Y on $\bar{M}$. From (1.19) and (1.20) we have

Theorem 1.2. The contact Bochner curvature tensor $\bar{B}$ of $\bar{M}$ vanishes if and only if the Bochner curvature tensor $\bar{B}'$ of $\bar{N}$ vanishes.

§2. Submanifolds and Riemannian fibre bundles

Let $\bar{M}$ and $\bar{N}$ be those of §1. Let M be an (n+1)-dimensional submanifold immersed in $\bar{M}$ and N be an n-dimensional submanifold immersed in $\bar{N}$. In the following we assume that M is tangent to the structure vector field ξ of $\bar{M}$ and there exists a fibration $\pi : M \longrightarrow N$ such that the diagram

(2.1) $$\begin{array}{ccc} M & \xrightarrow{\ i\ } & \bar{M} \\ {\scriptstyle \pi}\downarrow & & \downarrow{\scriptstyle \bar{\pi}} \\ N & \xrightarrow{\ i'\ } & \bar{N} \end{array}$$

commutes and the immersion i is a diffeomorphism on the fibres. Let g and G be the induced metric tensor fields of M and N respectively. Let ∇ (resp. ∇') be the operator of covariant differentiation with respect to g (resp. G). We denote by B (resp. B') in the sequel the second fundamental form of the immersion i (resp. i') and the associated second fundamental forms of B and B' will be denoted by A and A' respectively. For any vector fields X and Y on N, the Gauss formulas are given by

(2.2) $$\bar{\nabla}'_X Y = \nabla'_X Y + B'(X,Y) \quad \text{and} \quad \bar{\nabla}_{X^*} Y^* = \nabla_{X^*} Y^* + B(X^*,Y^*).$$

From (1.2) and (2.2) we have

$$(\nabla'_X Y)^* + (B'(X,Y))^* = -\phi^2 \nabla_{X^*} Y^* - \phi^2 B(X^*,Y^*)$$

$$= -\phi^2\nabla_{X*}Y* + B(X*,Y*).$$

Comparing the tangential and normal parts of this equation, we have respectively

$$(\nabla'_X Y)* = - \phi^2\nabla_{X*}Y*, \tag{2.3}$$

and

$$(B'(X,Y))* = B(X*,Y*). \tag{2.4}$$

Let D and D' be the operators of covariant differentiation with respect to the linear connections induced in the normal bundles of M and N respectively. For any tangent vector field X and any normal vector field V to N, we obtain

$$\bar{\nabla}'_X V = -A'_V X + D'_X V \quad \text{and} \quad \bar{\nabla}_{X*}V* = -A_{V*}X* + D_{X*}V*. \tag{2.5}$$

From (1.2) and (2.5) we see that

$$- (A'_V X)* + (D'_X V)* = \phi^2 A_{V*}X* - \phi^2 D_{X*}V*$$

$$= \phi^2 A_{V*}X* + D_{X*}V*.$$

Comparing the tangential and normal parts of this equation, we have respectively

$$(A'_V X)* = - \phi^2 A_{V*}X*, \tag{2.6}$$

and

$$(D'_X V)* = D_{X*}V*. \tag{2.7}$$

Proposition 2.1. M is a minimal submanifold if and only if N is a minimal submanifold.

Proof. Since the structure vector field ξ of $\bar{M}$ is tangent to M, we have

$$0 = \phi\xi = \bar{\nabla}_\xi \xi = \nabla_\xi \xi + B(\xi,\xi),$$

from which

$$B(\xi,\xi) = 0. \tag{2.8}$$

Now we take an orthonormal frame $e_1,\ldots,e_n$ for $T_{\pi(x)}(N)$. Then $e_1{}^*,\ldots,e_n{}^*$, ξ is an orthonormal frame for $T_x(M)$. Let m and m' be the mean curvature vectors of M and N respectively. Then (2.4) and (2.8) imply

$$(m')^* = \sum_{i=1}^{n} (B'(e_i,e_i))^* = \sum_{i=1}^{n} B(e_i{}^*,e_i{}^*) + B(\xi,\xi) = m,$$

that is,

$$(m')^* = m. \tag{2.9}$$

Therefore M is minimal if and only if N is minimal.

From (1.1) we have

Proposition 2.2. M is an anti-invariant submanifold of $\bar{M}$ if and only if N is an anti-invariant submanifold of $\bar{N}$.

Proposition 2.3 (Harada [14]). M is an invariant submanifold of $\bar{M}$ if and only if N is an invariant submanifold of $\bar{N}$.

From (2.7) and (2.9) we have $(D'_X m')^* = D_{X^*} m$. Consequently, we have

Proposition 2.4. If the mean curvature vector m of M is parallel, then the mean curvature vector m' of N is also parallel.

By (2.4) we have

Proposition 2.5. If M is totally geodesic in $\bar{M}$, then N is totally geodesic in $\bar{N}$.

Proposition 2.6. Let M and N be invariant submanifolds. Then M is totally geodesic if and only if N is totally geodesic.

Proof. If M is invariant, we have $B(X,\xi) = 0$ for any vector field X on M (see §6 of Chapter II). From this and (2.4) we have our proposition.

§3. Contact totally umbilical submanifolds

Let M be a submanifold tangent to the structure vector field ξ of a Sasakian manifold $\bar{M}$. If the second fundamental form B of M is of the form

$$(3.1) \qquad B(X,Y) = [g(X,Y) - \eta(X)\eta(Y)]\alpha + \eta(X)B(Y,\xi) + \eta(Y)B(X,\xi)$$

for any vector fields X and Y tangent to M, where α denotes a normal vector field to M, then M is said to be contact totally umbilical.

The purpose of this section is to prove the following

Theorem 3.1 (Kon [31]). Let $\bar{M}$ be a regular Sasakian manifold and M be an immersed submanifold of $\bar{M}$ tangent to the structure vector field ξ of $\bar{M}$. Suppose that there is a fibration $\pi: M \longrightarrow N$ where N is an immersed submanifold of $\bar{N} = \bar{M}/\xi$ such that the diagram

$$\begin{array}{ccc} M & \xrightarrow{\ \ i\ \ } & \bar{M} \\ {\scriptstyle\pi}\downarrow & & \downarrow{\scriptstyle\bar{\pi}} \\ N & \xrightarrow{\ \ i'\ \ } & \bar{N} \end{array}$$

commutes and the immersion i is a diffeomorphism on the fibres. Then M is contact totally umbilical if and only if N is totally umbilical.

Proof. Let $e_1,\dots,e_n$ be an orthonormal frame for $T_{\pi(x)}(N)$. We put $\beta' = \frac{1}{n}\Sigma B'(e_i,e_i)$. If N is totally umbilical in $\bar{N}$, then $B'(X,Y) = G(X,Y)\beta'$

for any vectors X, Y on N. Then (2.4) implies that $B(X^*,Y^*) = g(X^*,Y^*)\beta$, where $\beta = (\beta')^*$. We have seen that $B(\xi,\xi) = 0$. On the other hand, the horizontal space of π is given by $\{\phi^2 X : X \in T_x(M)\}$. Consequently, if N is totally umbilical, we have

$$B(X,Y) = [g(X,Y) - \eta(X)\eta(Y)]\beta + \eta(X)B(Y,\xi) + \eta(Y)B(X,\xi)$$

for any vector fields X and Y on M. Therefore M is contact totally umbilical. Conversely, if M is contact totally umbilical, by (3.1), we have $B(\phi^2 X,\phi^2 Y) = g(\phi^2 X,\phi^2 Y)$. From this and (2.4) we see that N is totally umbilical.

Remark. If X, Y are orthogonal to ξ, then (3.1) becomes $B(X,Y) = g(X,Y)\alpha$. Therefore the notion of contact totally umbilical submanifold is equivalent to that of H-totally umbilical anti-invariant submanifolds in §8 of Chapter IV.

Let M be an anti-invariant submanifold tangent to the structure vector field ξ of a Sasakian manifold $\bar{M}$. If M is contact totally umbilical, then we have

$$B(X,Y) = [g(X,Y) - \eta(X)\eta(Y)]\alpha + \eta(X)\phi Y + \eta(Y)\phi X. \tag{3.2}$$

Conversely, if the second fundamental form B of M is of the form (3.2), then, putting $Y = \xi$ in (3.2), we have $B(X,\xi) = \phi X$, which means that M is anti-invariant.

Remark. If a submanifold M is contact totally umbilical and minimal in $\bar{M}$, then we see that $B(\phi^2 X,\phi^2 Y) = 0$ for any vector fields X, Y tangent to M. Thus we have H = 0, H being defined in §1 of Chapter IV.

§4. Properties of the second fundamental forms

Let M be an (n+1)-dimensional submanifold of a (2m+1)-dimensional Sasakian manifold $\bar{M}$ and N be an n-dimensional submanifold of a real 2m-dimensional Kaehlerian manifold $\bar{N}$ such that the diagram (2.1) commutes and the immersion i is a diffeomorphism on the fibres. We denote by S and S' the square of the length of the second fundamental forms of M and N respectively. Then, by (1.1) and (2.4), we have

$$\begin{aligned} S &= \sum_{i,j} \bar{g}(B(e_i^*,e_j^*),B(e_i^*,e_j^*)) + 2\sum_i \bar{g}(B(e_i^*,\xi),B(e_i^*,\xi)) \\ &= \sum_{i,j} \bar{G}(B'(e_i,e_j),B'(e_i,e_j)) + 2\sum_i \bar{g}((\phi e_i^*)^{\perp},(\phi e_i^*)^{\perp}) \\ &= S' + 2\sum_i \bar{G}((Je_i)^{\perp},(Je_i)^{\perp}), \end{aligned}$$

where we have used the fact that $\phi e_i^* = \nabla_{e_i^*}\xi + B(e_i^*,\xi)$ and $\perp$ denotes the normal projection. Thus we have

$$S = S' + 2\sum_i \bar{G}((Je_i)^{\perp},(Je_i)^{\perp}). \tag{4.1}$$

Hence if S = S', we have $(Je_i)^{\perp} = 0$, which shows that N is invariant and consequently M is invariant by Proposition 2.3. Conversely, if N is invariant, then S = S'. Therefore we have

Proposition 4.1. M and N are invariant submanifold if and only if S = S'.

Since $\bar{G}((Je_i)^{\perp},(Je_i)^{\perp}) \leq 1$ for all i (=1,...,n), we have the following inequality

$$S \leq S' + 2n. \tag{4.2}$$

If the equality in (4.2) holds, we see that $(Je_i)^{\perp} = Je_i$ for all i, in which case N is an anti-invariant submanifold. From this and Proposition 2.2 we have

Proposition 4.2. M and N are anti-invariant submanifolds if and only if $S = S' + 2n$.

Corollary 4.1. Let M and N be anti-invariant submanifolds of $\bar{M}$ and $\bar{N}$ respectively. Then N is totally geodesic if and only if $S = 2n$.

Example 4.1. Let CP^n be a complex projective space of constant holomorphic sectional curvature 4 and RP^n be a real projective space of constant curvature 1. We denote by (RP^n,S^1) the circle bundle over RP^n. Now we consider the following commutative diagram:

$$\begin{array}{ccc} (RP^n,S^1) & \longrightarrow & S^{2n+1} \\ \downarrow & & \downarrow \\ RP^n & \longrightarrow & CP^n \end{array} .$$

The real projective space RP^n is a totally geodesic anti-invariant submanifold of CP^n. Thus (RP^n,S^1) is an anti-invariant submanifold of S^{2n+1} with $S = 2n$. Therefore (RP^n,S^1) is an example of submanifold for which $T = 0$ (see 4 of Chapter IV).

Lemma 4.1. Let M and N be anti-invariant submanifolds. Then the Riemannian curvature tensors R and R' of M and N satisfy

$$(R'(X,Y)Z)^* = R(X^*,Y^*)Z^*. \tag{4.3}$$

Proof. From (2.3) we have

$$(\nabla'_X Y)^* = \nabla_{X^*}Y^* - \eta(\nabla_{X^*}Y^*)\xi.$$

On the other hand, from Proposition 7.2 of Chapter II, the structure vector field ξ of $\bar{M}$ is parallel on M. Thus we have

$$\eta(\nabla_{X^*}Y^*) = \nabla_{X^*}g(Y^*,\xi) - g(Y^*,\nabla_{X^*}\xi) = 0.$$

Therefore we have

$$(4.4) \qquad (\nabla'_X Y)^* = \nabla_{X^*}Y^*,$$

from which we see that

$$\begin{aligned}(R'(X,Y)Z)^* &= (\nabla'_X\nabla'_Y Z - \nabla'_Y\nabla'_X Z - \nabla'_{[X,Y]}Z)^* \\ &= \nabla_{X^*}\nabla_{Y^*}Z^* - \nabla_{Y^*}\nabla_{X^*}Z^* - \nabla_{[X^*,Y^*]}Z^* \\ &= R(X^*,Y^*)Z^*.\end{aligned}$$

From Proposition 7.2 of Chapter II and Lemma 4.1, we have

Theorem 4.1. Let M and N be anti-invariant submanifolds of $\bar{M}$ and $\bar{N}$ respectively. Then M is flat if and only if N is flat.

From Lemma 5.1 of Chapter III, Proposition 2.2 of Chapter IV and Theorem 4.1 we have

Corollary 4.2. Let M be an (n+1)-dimensional anti-invariant submanifold of a (2n+1)-dimensional Sasakian manifold $\bar{M}$ and let N be an n-dimensional anti-invariant submanifold of a real 2n-dimensional Kaehlerian manifold $\bar{N}$. Then the normal connection of M is flat if and only if the normal connection of N is flat.

§5. Submanifolds of CP^m

First of all, we remember the following theorem due to Simons [51]:

Theorem 5.1. Let M be an n-dimensional compact orientable minimal submanifold of S^{n+p}. If the square of the length of the second fundamental form of

M satisfies

$$(5.1) \qquad S \leq \frac{n}{2 - \frac{1}{p}},$$

then either M is totally geodesic or $S = n/(2 - \frac{1}{p})$.

Moreover, we need the following theorem of Chern-do Carmo-Kobayashi [10]:

Theorem 5.2. The Veronese surface in S^4 and the generalized Clifford hypersurface $M_{p,q}$ are the only compact orientable minimal submanifolds of dimension n in S^{n+p} satisfying $S = n/(2 - \frac{1}{p})$.

For the generalized Clifford hypersurface $M_{p,q}$, see Example 6.1 in §6 of this chapter.

Now we consider the following commutative diagram similar to that in the previous sections:

$$\begin{array}{ccc} M^{n+1} & \xrightarrow{\quad i \quad} & S^{2m+1} \\ \pi \downarrow & & \downarrow \bar{\pi} \\ N^n & \xrightarrow{\quad i' \quad} & CP^m \end{array} .$$

We put $p = 2m - n$, p being the real codimension of M in $\bar{M}$ and at the same time that of N in $\bar{N}$. Now we prove a pinching theorem of Okumura [48].

Theorem 5.3. If an n-dimensional compact orientable minimal submanifold N of a complex projective space CP^m satisfies

$$(5.2) \qquad S' < \frac{n+3-4p}{2 - \frac{1}{p}},$$

then N is a totally geodesic complex projective space $CP^{n/2}$.

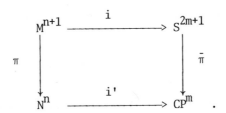

Proof. First of all, from (4.1), we have

$$(5.3) \qquad S \leq S' + 2p, \qquad p = 2m - n.$$

Thus from the assumption (5.2) and (5.3) we obtain

$$(5.4) \qquad S < \frac{n+3-4p}{2 - \frac{1}{p}} + 2p = \frac{n+1}{2-\frac{1}{p}}.$$

On the other hand, by Proposition 2.1, M is a compact minimal submanifold of S^{2m+1}. Thus applying Theorem 5.1, we see that M is a totally geodesic submanifold S^{n+1} of S^{2m+1}. Hence, Proposition 2.5 implies that N is totally geodesic in CP^m. Therefore N is a real projective space RP^n or a complex projective space $CP^{n/2}$. But, from the assumption, we have $n+3-4p > 0$, that is, $(n+3)/4 > p$. Therefore we have $n > p$, which means that N is not anti-invariant in CP^m. Thus we see that N must be a complex projective space $CP^{n/2}$.

Let N be an n-dimensional minimal submanifold of CP^m. Then from the Gauss equation (2.4) of Chapter II, we have

$$(5.5) \qquad \bar{G}(B'(X,Y),B'(X,Y)) - \bar{G}(B'(X,X),B'(Y,Y)) = \bar{K}'(X,Y) - K'(X,Y)$$

where $\bar{K}'(X,Y)$ and $K'(X,Y)$ are sectional curvatures of CP^m and N respectively. From the minimality of N and (5.5) we have

$$(5.6) \qquad 0 \leq S' = \sum_{i,j=1}^{n} \bar{K}'(e_i,e_j) - r',$$

where r' denotes the scalar curvature of N. On the other hand, by (6.2) of Chapter I, we have

$$(5.7) \qquad \sum_{i,j=1}^{n} \bar{K}'(e_i,e_j) = 3 \sum_{i,j=1}^{n} \bar{G}(e_i,Je_j)^2 + n(n-1).$$

Theorem 5.4. Let N be an n-dimensional compact orientable minimal submanifold of CP^m with scalar curvature r'. If r' satisfies the inequality

$$(5.8) \qquad r' \geq n(n+2) - \frac{n+1}{2-\frac{1}{p}},$$

then N is a totally geodesic complex projective space $CP^{n/2}$.

Proof. From (4.1), (5.6) and (5.7) we have

$$(5.9) \qquad S = n(n-1) - r' + 3\sum_{i,j=1}^{n} \bar{G}(e_i, Je_j)^2 + 2\sum_{i=1}^{n} \bar{G}((Je_i)^{\perp}, (Je_i)^{\perp}).$$

Since we have

$$3\sum_{i,j=1}^{n} \bar{G}(e_i, Je_j)^2 = 3\sum_{i=1}^{n} \bar{G}((Je_i)^T, (Je_i)^T),$$

where $(Je_i)^T$ is the tangential part of Je_i, equation (5.9) becomes

$$(5.10) \qquad S = n(n-1) - r' + 3n - \sum_{i=1}^{n} \bar{G}((Je_i)^{\perp}, (Je_i)^{\perp})$$

$$\leq n(n+2) - r'.$$

From the assumption (5.8) and (5.10) we obtain the inequality

$$(5.11) \qquad S \leq \frac{n+1}{2-\frac{1}{p}}.$$

Therefore Theorem 5.1 implies that S = 0 or $S = (n+1)/(2-\frac{1}{p})$. If M is not totally geodesic, it is a Veronese surface in S^4 by Theorem 5.2. Since the ambient manifold S^{2m+1} is odd dimensional, this is not possible. Therefore M is a totally geodesic S^{n+1} of S^{2m+1}. Thus N is a totally geodesic complex projective space $CP^{n/2}$.

Remark. Theorem 5.4 is essentially due to Lawson [34]. He proved the theorem under the assumption that $\dim N = n$ = even.

From (5.9) and (5.10) we have

$$S = n(n+2) - r' - \sum_{i=1}^{n} \bar{G}((Je_i)^{\perp},(Je_i)^{\perp}), \tag{5.12}$$

$$S = n(n+1) - r' + \sum_{i=1}^{n} \bar{G}((Je_i)^{T},(Je_i)^{T}). \tag{5.13}$$

On the other hand, the scalar curvature r of M satisfies

$$S = n(n+1) - r. \tag{5.14}$$

From (5.12), (5.13) and (5.14) we obtain the equations:

$$r' - (r + n) = - \sum_{i=1}^{n} \bar{G}((Je_i)^{\perp},(Je_i)^{\perp}), \tag{5.15}$$

$$r' - r = \sum_{i=1}^{n} \bar{G}((Je_i)^{T},(Je_i)^{T}), \tag{5.16}$$

from which we have

Proposition 5.1. Let M be an (n+1)-dimensional minimal submanifold of S^{2m+1} and N be an n-dimensional minimal submanifold of CP^m. Then the scalar curvatures r of M and r' of N satisfy

$$r + n \geq r', \tag{5.17}$$

and the equality holds if and only if M and N are both invariant submanifolds of S^{2m+1} and CP^m respectively.

Proposition 5.2. Let M and N be as in Proposition 5.1. Then the scalar curvature r and r' satisfy

$$r \leq r', \tag{5.18}$$

and the equality holds if and only if M and N are both anti-invariant submanifolds of S^{2m+1} and CP^m respectively.

§6. Real hypersurfaces of CP^m

Under the same situation as in §5, we study a real hypersurfaces N of dimension 2m-1 of a real 2m-dimensional complex projective space CP^m.

Let v be a unit normal vector field of N in CP^m. It is possible to choose an orthonormal frame field on N of the form: $e_1, Je_1, \ldots, e_{m-1}, Je_{m-1}$, Jv, where J denotes the almost complex structure of CP^m. Then equation (4.1) becomes

$$(6.1) \qquad S = S' + 2.$$

Now we state an example of a real hypersurface of CP^m given by Lawson [34].

Example 6.1. In S^{2m+1} we have the family of generalized Clifford surfaces

$$M_{p,q} = S^p\left(\sqrt{\frac{p}{2m}}\right) \times S^q\left(\sqrt{\frac{q}{2m}}\right),$$

where p + q = 2m. By choosing the spheres in such a way that they lie in complex subspaces we get a fibration

$$M_{2p+1,2q+1} \longrightarrow M^C_{p,q}$$

compatible with

$$S^{2m+1} \longrightarrow CP^m,$$

where p + q = m - 1. Whenever p = 0, this hypersurface is homogeneous, positively curved manifold diffeomorphic to the sphere. $M^C_{p,q}$ is a minimal hypersurface of CP^m.

Theorem 6.1 (Lawson [34]). Let N be a compact orientable minimal hypersurface of CP^m. If $S' \leq 2(m-1)$, then $S' = 2(m-1)$ and $N = M^C_{p,q}$ for some p, q.

Proof. Using (6.1), we have

$$S \leq 2m. \tag{6.2}$$

From Theorem 5.1 and (6.2), we obtain S = 0, which means that M is totally geodesic in S^{2m+1} and hence N is totally geodesic in CP^m and is a complex submanifold of CP^m, or S = 2m. But the real codimension of N is 1, and hence N is not a complex submanifold. Therefore we have S = 2m. Thus we apply Theorem 5.2, and see that M^{2m} is $M_{2p+1,2q+1}$ for some p, q. Therefore N is $M^C_{p,q}$ for some p, q.

From (5.6) we have

$$S' = (2m-1)(2m+3) - r'. \tag{6.3}$$

From Theorem 6.1 and (6.3) we have

Corollary 6.1. Let N be a compact orientable minimal hypersurface of CP^m. If $r' \geq 2(m+1)(2m-1)$, then $r' = 2(m+1)(2m-1)$ and $N = M^C_{p,q}$ for some p, q.

As a local version of this, we have

Theorem 6.2 (Lawson [34]). Let N be a minimal hypersurface of CP^m having the scalar curvature $r' = 2(m+1)(2m-1)$. Then N is an open submanifold of $M^C_{p,q}$ for some p, q.

§7. Anti-holomorphic submanifolds

Let N be an n-dimensional submanifold of a real 2m-dimensional Kaehlerian manifold $\bar{N}$. If $T_x(N)^{\perp} \perp JT_x(N)^{\perp}$ for all $x \in N$, then N is called an anti-holomorphic submanifold of $\bar{N}$. Such a submanifold has been studied by Okumura [48]. In the following, we study an n-dimensional anti-holomorphic submanifold N of a real 2m-dimensional complex projective space CP^m compatible with $\bar{\pi} : S^{2m+1} \longrightarrow CP^m$ and prove a theorem of Okumura [48]. From the definition of anti-holomorphic submanifold, we have $n \geq p$, where $p = 2m - n$.

Theorem 7.1. Let N be an n-dimensional compact orientable minimal anti-holomorphic submanifold of a complex projective space CP^m. If

$$(7.1) \qquad S' \leq \frac{n+3-4p}{2 - \frac{1}{p}}, \qquad p = 2m - n,$$

then N is $M^C_{p,q}$ in $CP^{(n+1)/2}$.

Proof. Since N is anti-holomorphic, we have

$$(7.2) \qquad \sum_i \bar{G}((Je_i)^{\perp}, (Je_i)^{\perp}) = p.$$

From (4.1) and (7.2) we have

$$(7.3) \qquad S = S' + 2p.$$

Using the assumption (7.1), we obtain the inequality

$$(7.4) \qquad S \leq \frac{n + 1}{2 - \frac{1}{p}}.$$

Therefore, Theorem 5.1 implies that $S = 0$ or $S = (n+1)/(2-\frac{1}{p})$. If $S = 0$, then M is a sphere S^{n+1} and hence N is a complex projective space. But N being anti-holomorphic, this is not possible. When $S = (n+1)/(2-\frac{1}{p})$, we know by Theorem 5.2 that M is $M_{2p+1,2q+1}$ in S^{n+2} because M is compatible with the fibration. Therefore we have $N = M^C_{p,q}$ in $CP^{(n+1)/2}$.

Remark. Any real hypersurface N of CP^m is an anti-holomorphic submanifold of CP^m. Thus, putting p = 1 in Theorem 7.1, we have Theorem 6.1.

§8. Totally umbilical submanifolds of complex space forms

First of all, we prove

Theorem 8.1 (Chen-Ogiue [7]). Let N be an n-dimensional ($n \geq 2$) totally umbilical submanifold of a real 2m-dimensional complex space form $\bar{N}$ of constant holomorphic sectional curvature $c \neq 0$. Then N is one of the following submanifolds:

(a) a complex space form immersed in $\bar{N}$ as a totally geodesic submanifold, or

(b) a real space form immersed in $\bar{N}$ as an anti-invariant totally geodesic submanifold, or

(c) a real space form immersed in $\bar{N}$ as an anti-invariant submanifold with nonzero parallel mean curvature vector.

Case (b) occurs only when $m \geq n$ and case (c) occurs only when $m > n$.

Proof. Let N be a totally umbilical submanifold of $\bar{N}^m(c)$ ($c \neq 0$). Then we have

$$B'(X,Y) = G(X,Y)m', \tag{8.1}$$

where B' and m' denote the second fundamental form and the mean curvature vector of N respectively. From (2.1) of Chapter II and (8.1) we have

$$(\nabla_X B')(Y,Z) = G(Y,Z)D'_X m', \tag{8.2}$$

from which, using the Codazzi equation (2.5) of Chapter II,

(8.3) $$(\bar{R}'(X,Y)Z)^{\perp} = G(Y,Z)D'_X m' - G(X,Z)D'_Y m'.$$

If $\dim N \geq 3$, for any vector field X tangent to N we can choose a unit vector field Y tangent to N and orthogonal to X and JX. For such a choice it follows from (8.3) that

(8.4) $$(\bar{R}'(X,Y)Y)^{\perp} = D'_X m'.$$

On the other hand, (6.2) of Chapter I implies that $(\bar{R}'(X,Y)Y)^{\perp} = 0$ so that $D'_X m' = 0$ for any vector field X tangent to N, which means that the mean curvature vector m' of N is parallel. Therefore, by (8.3) we have

(8.5) $$(\bar{R}'(X,Y)Z)^{\perp} = 0$$

for any vector fields X, Y and Z tangent to N.

If $\dim N = 2$, put $N = N_1 \cup N_2$, where $N_1 = \{x \in N : JT_x(N) = T_x(N)\}$ and $N_2 = \{x \in N : JT_x(N) \neq T_x(N)\}$. Then we can see that N_2 is an open submanifold of N, where the preceding argument is available and consequently (8.5) holds. Let N_1' be the set of all interior points of N_1. Then N_1' is a complex analytic submanifold of $\bar{N}$ so that the mean curvature vector $m' = 0$ on N_1', and hence (8.5) holds on N_1'. Since (8.5) is a tensorial equation, this holds on N.

Thus (8.5) holds for N with $\dim N \geq 2$. From Proposition 5.2 of Chapter II and (8.5), we see that N is either a complex submanifold of $\bar{N}$ or an anti-invariant submanifold of $\bar{N}$. If N is a complex submanifold of $\bar{N}$, N being minimal, it is a totally geodesic submanifold. If N is an anti-invariant submanifold of $\bar{N}$, then (1.5) of Chapter III and (8.1) show that N is a real space form of constant curvature $\frac{1}{4}c + \bar{G}(m',m')$. Since N is anti-invariant, we have $m \geq n$. Moreover, if $m' \neq 0$, then $m > n$ holds. In fact, since $D'_X m' = 0$, from (2.6) of Chapter II and (8.1) we have

$$\bar{G}(\bar{R}'(X,Y)m',JY) = 0$$

for any vector fields X, Y tangent to N. Thus (6.2) of Chapter I implies that

$$\bar{G}(JY,m')\bar{G}(JX,JY) = \bar{G}(JX,m')\bar{G}(JY,JY).$$

Choose Y in such a way that $\bar{G}(JY,m') = 0$. Then we have

$$\bar{G}(JX,m')\bar{G}(Y,Y) = 0,$$

which implies that m' is perpendicular to $JT_x(N)$. Hence we have $m > n$. These considerations prove our assertion.

Remark. A totally umbilical submanifold of a real projective space $RP^n(\frac{1}{4}c)$ can be imbedded in a complex projective space $CP^n(c)$ as an anti-invariant, totally umbilical submanifold. In particular, if N is not totally geodesic in $RP^n(\frac{1}{4}c)$, then N is not totally geodesic in $CP^n(c)$.

In the following, we consider an (n+1)-dimensional contact totally umbilical submanifold M of a (2m+1)-dimensional Sasakian space form of constant ϕ-sectional curvature $c \neq -3$ compatible with the fibration $\bar{\pi} : \bar{M} \longrightarrow \bar{N}$ as in §2. If M is contact totally umbilical, then by Theorem 3.1, N is totally umbilical. Thus Propositions 2.2, 2.3 and Theorem 8.1 show that M is either an invariant submanifold of $\bar{M}$ or an anti-invariant submanifold of $\bar{M}$.

If M is an invariant submanifold of $\bar{M}$, then by Proposition 2.6 and (a) of Theorem 8.1, M is totally geodesic in $\bar{M}$. Let M be an anti-invariant submanifold of $\bar{M}$. If N is totally geodesic, anti-invariant in $\bar{N}$, then Corollary 4.1 shows that $S = 2n$, in which case $B(X^*,Y^*) = 0$ for any vector fields X, Y on N. If N is an anti-invariant submanifold with nonzero parallel mean curvature vector, then (2.7) and (2.9) imply that the mean curvature vector m of M satisfies $D_{X*}m = 0$, which means that m is parallel with respect to the horizontal direction. From these considerations we have

Theorem 8.2. Let M be an (n+1)-dimensional ($n \geq 2$) contact totally umbilical submanifold of a Sasakian space form $\bar{M}$ of constant ϕ-sectional curvature $c \neq -3$. Then M is one of the following submanifolds:

(a) an invariant Sasakian space form immersed in $\bar{M}$ as a totally geodesic submanifold, or

(b) an anti-invariant minimal submanifold immersed in $\bar{M}$ with $S = 2n$, or

(c) an anti-invariant submanifold with nonzero mean curvature vector m such that $D_{X*}m = 0$.

Case (b) occurs only when $m \geq n$ and case (c) occurs only when $m > n$.

Remark. Case (b) in Theorem 8.2 states that M is locally a Riemannian direct product $M^n \times M^1$, where M^n is an anti-invariant totally geodesic real space form of $\bar{M}$ and M^1 is a 1-dimensional space generated by ξ. Case (c) in Theorem 8.2 states that M is locally a Riemannian direct product $M^n \times M^1$, where M^n is an anti-invariant real space form with nonzero parallel mean curvature vector and M^1 is a 1-dimensional space generated by ξ. (See also Ishihara-Kon [16].)

BIBLIOGRAPHY

This bibliography contains not only the papers and books quoted in the text but also recent papers on invariant submanifolds of Kaehlerian and Sasakian manifolds.

[1] K. Abe, Applications of Riccati type differential equation to Riemannian manifolds with totally geodesic distribution, Tôhoku Math. J., 25 (1973), 425-444.

[2] R. L. Bishop and R. J. Crittenden, Geometry of manifolds, Academic Press, New York, 1964.

[3] D. E. Blair, On the geometric meaning of the Bochner tensor, Geometria Dedicata, 4 (1975), 33-38.

[4] D. E. Blair and K. Ogiue, Geometry of integral submanifolds of a contact distribution, Illinois J. Math., 19 (1975), 269-276.

[5] B. Y. Chen, Geometry of submanifolds, Marcel Dekker, Inc., New York, 1973.

[6] B. Y. Chen and K. Ogiue, On totally real submanifolds, Trans. Amer. Math. Soc., 193 (1974), 257-266.

[7] _____, Two theorems on Kaehler manifolds, Michigan Math. J., 21 (1974), 225-229.

[8] S. S. Chern, Einstein hypersurfaces in a Kaehler manifold of constant holomorphic curvature, J. Differential Geometry, 1. (1967), 21-31.

[9] _____, Minimal submanifolds in a Riemannian manifold, Univ. of Kansas, Technical Report 19, 1968.

[10] S. S. Chern, M. do Carmo and S. Kobayashi, Minimal submanifolds with second fundamental form of constant length, Functional analysis and related fields, Springer, (1970), 59-75.

[11] J. Erbacher, Reduction of the codimension of an isometric immersion, J. Differential Geometry, 5 (1971), 333-340.

[12] _____, Isometric immersions of constant mean curvature and triviality of the normal connection, Nagoya Math. J., 45 (1971), 139- 165.

[13] S. I. Goldberg, Curvature and homology, Academic Press, New York, 1962.

[14] M. Harada, On Sasakian submanifolds, Tôhoku Math. J., 25 (1973), 103-109.

[15] C. S. Houh, Some totally real minimal surfaces in CP^2, Proc. Amer. Math. Soc., 40 (1973), 240-244.

[16] I. Ishihara and M. Kon, Contact totally umbilical submanifolds of a Sasakian space form, to appear.

[17] S. Ishihara and M. Konishi, Differential geometry of fibred spaces, Study Group of Differential Geometry, Japan, 1973.

[18] K. Kenmotsu, Invariant submanifolds in a Sasakian manifold, Tôhoku Math. J., 21 (1969), 495-500.

[19] _____, Local classification of invariant η-Einstein submanifolds of codimension 2 in a Sasakian manifold with constant ϕ-sectional curvature, Tôhoku Math. J., 22 (1970), 270-272.

[20] S. Kobayashi, Hypersurfaces of complex projective space with constant scalar curvature, J. Differential Geometry, 1 (1967), 369-370.

[21] S. Kobayashi and K. Nomizu, Foundations of differential geometry, Vol. I and II, Interscience Publishers, 1963 and 1969.

[22] M. Kon, Invariant submanifolds of normal contact metric manifolds, Kōdai Math. Sem. Rep., 25 (1973), 330-336.

[23] _____, Kaehler immersions with trivial normal connection, TRU Math., 9 (1973), 29-33.

[24] _____, On some invariant submanifolds of normal contact metric manifolds, Tensor N. S., 28 (1974), 133-138.

[25] _____, Totally real minimal submanifolds with parallel second fundamental form, Atti della Accademia Nazionale dei Lincei, 57 (1974), 187-189.

[26] _____, On some complex submanifolds in Kaehler manifolds, Canad. J. Math., 26 (1974), 1442-1449.

[27] _____, Complex submanifolds with constant scalar curvature in a Kaehler manifold, J. Math. Soc. Japan, 27 (1975), 76-81.

[28] _____, Invariant submanifolds in Sasakian manifolds, Math. Ann., 219 (1976), 277-290

[29] _____, Totally real submanifolds in a Kaehler manifold, to appear in J. Differential Geometry.

[30] M. Kon, Kaehler immersions with vanishing Bochner curvature tensors, Kōdai Math. Sem. Rep., 27 (1976), 329-333.

[31] _____, Remarks on anti-invariant submanifolds of a Sasakian manifold, to appear in Tensor N. S.

[32] M. Kon and T. Ikawa, Sasakian manifolds with vanishing contact Bochner curvature tensor and constant scalar curvature, to appear in Colloquium Math.

[33] H. F. Lai, Characteristic classes of real manifolds immersed in complex manifolds, Trans. Amer. Math. Soc., 172 (1972), 1-33.

[34] H. B. Lawson, Jr., Local rigidity theorems for minimal hypersurfaces, Ann. of Math., 89 (1969), 187-197.

[35] _____, Rigidity theorems in rank-1 symmetric spaces, J. Differential Geometry, 4 (1970), 349-357.

[36] G. D. Ludden, M. Okumura and K. Yano, A totally real surface in CP^2 that is not totally geodesic, Proc. Amer. Math. Soc., 53 (1975), 186-190.

[37] _____, Totally real submanifolds of complex manifolds, Atti della Accademia Nazionale dei Lincei, 58 (1975), 346-353.

[38] _____, Anti-invariant submanifolds of almost contact metric manifolds, to appear in Math. Ann.

[39] M. Matsumoto and G. Chūman, On the C-Bochner curvature tensor, TRU Math., 5 (1969), 21-30.

[40] M. Matsumoto and S. Tanno, Kaehlerian spaces with parallel or vanishing Bochner curvature tensor, Tensor N. S., 27 (1973), 291-294.

[41] H. Nakagawa, Einstein Kaehler manifolds immersed in a complex projective space, Canad. J. Math., 28 (1976), 1-8.

[42] H. Nakagawa and K. Ogiue, Certain Kaehler submanifolds immersed in Kaehler manifolds of constant holomorphic curvature, to appear in Trans. Amer. Math. Soc.

[43] H. Nakagawa and R. Takagi, On locally symmetric Kaehler submanifolds in a complex projective space, to appear.

[44] K. Nomizu and B. Smyth, Differential geometry of complex hypersurfaces II, J. Math. Soc. Japan, 20 (1968), 498-521.

[45] K. Ogiue, Differential geometry of Kaehler submanifolds, Advances in Math., 13 (1974), 73-114.

[46] M. Okumura, On infinitesimal conformal and projective transformations of normal contact spaces, Tôhoku Math. J., 14 (1962), 398-412.

[47] _____, On some real hypersurfaces of a complex projective space, Trans. Amer. Math. Soc., 212 (1975), 355-364.

[48] _____, Submanifolds of real codimension p of a complex projective space, Atti della Accademia Nazionale dei Lincei, 58 (1975), 544-555.

[49] B. O'Neill, The fundamental equations of a submersion, Michigan Math. J., 13 (1966), 459-469.

[50] S. Sasaki, Almost contact manifolds, Lecture Notes, Tôhoku University, 1965.

[51] J. Simons, Minimal varieties in riemannian manifolds, Ann. of Math., 88 (1968), 62-105.

[52] B. Smyth, Differential Geometry of complex hypersurfaces, Ann. of Math., 85 (1967), 247-266.

[53] _____, Homogeneous complex hypersurfaces, J. Math. Soc. Japan, 20 (1968), 643-647.

[54] _____, Submanifolds of constant mean curvature, Math. Ann., 205 (1973), 265-280.

[55] S. Tachibana, On the Bochner curvature tensor, Nat. Sci. Rep. of Ochanomizu Univ., 18 (1967), 15-19.

[56] R. Takagi, On homogeneous real hypersurfaces in a complex projective space, Osaka J. Math., 10 (1973), 495-506.

[57] _____, Real hypersurfaces in a complex projective space with constant principal curvature, J. Math. Soc. Japan, 27 (1975), 43-53.

[58] T. Takahashi, Hypersurface with parallel Ricci tensor in a space of constant holomorphic sectional curvature, J. Math. Soc. Japan, 19 (1967), 199-204.

[59] S. Tanno, Sasakian manifolds with constant ϕ-holomorphic sectional curvature, Tôhoku Math. J., 21 (1969), 501-507.

[60] _____, Isometric immersions of Sasakian manifolds in spheres, Kōdai Math. Sem. Rep., 21 (1969), 448-458.

[61] _____, Compact complex submanifolds immersed in complex projective spaces, J. Differential Geometry, 8 (1973), 629-641.

[62] R. O. Wells, Jr., Compact real submanifolds of a complex manifold with nondegenerate holomorphic tangent bundles, Math. Ann., 179 (1969), 123-129.

[63] S. Yamaguchi, M. Kon and Y. Miyahara, A theorem on C-totally real minimal surface, Proc. Amer. Math. Soc., 54 (1976), 276-280.

[64] S. Yamaguchi, M. Kon and T. Ikawa, On C-totally real submanifolds, to appear in J. Differential Geometry.

[65] K. Yano, On n-dimensional Riemannian spaces admitting a group of motions of order $n(n-1)/2+1$, Trans. Amer. Math. Soc., 74 (1953), 260-279.

[66] _____, On a structure defined by a tensor field f of type (1,1) satisfying $f^3 + f = 0$, Tensor N. S., 14 (1963), 99-109.

[67] _____, Differential geometry on complex and almost complex spaces, Pergamen Press, New York, 1965.

[68] _____, Integral formulas in Riemannian geometry, Marcel Dekker, Inc., New York, 1970.

[69] _____, Totally real submanifolds of a Kaehlerian manifold, to appear in J. Differential Geometry.

[70] _____, Differential geometry of totally real submanifolds, to appear.

[71] _____, Anti-invariant submanifolds of a Sasakian manifold with vanishing contact Bochner curvature tensor, to appear.

[72] _____, Differential geometry of anti-invariant submanifolds of a Sasakian manifold, to appear.

[73] K. Yano and S. Bochner, Curvature and Betti numbers, Ann. Math. Studies, 32, 1953.

[74] K. Yano and S. Ishihara, Fibred spaces with invariant Riemannian metric, Kōdai Math. Sem. Rep., 19 (1967), 317-360.

[75] _____, Invariant submanifolds of almost contact manifolds, Kōdai Math. Sem. Rep., 21 (1969), 448-458.

[76] _____, Submanifolds with parallel mean curvature vector, J. Differential Geometry, 6 (1971), 95-118.

[77] K. Yano and M. Kon, Totally real submanifolds of complex space forms I, Tôhoku Math. J., 28 (1976), 215-225.

[78] _____, Totally real submanifolds of complex space forms II, Kōdai Math. Sem. Rep., 27 (1976), 385-399.

[79] _____, Anti-invariant submanifolds of Sasakian space forms I, to appear in Tôhoku Math. J.

[80] _____, Anti-invariant submanifolds of Sasakian space forms II, J. Korean Math. Soc., 13 (1976), 1-14.

[81] _____, Submanifolds of an even dimensional sphere, to appear in Geometria Dedicata.

[82] S. T. Yau, Submanifolds with constant mean curvature I, Amer. J. Math., 96 (1974), 346-366.

[83] _____, Submanifolds with constant mean curvature II, Amer. J. Math., 97 (1975), 76-100.

AUTHOR INDEX

SUBJECT INDEX

8460-77-9
5-10

8480-77-3
5-10